# Gotham Knights
## Complete Guide

Luca Kross

ISBN: 979-8-3597-2874-4



ISBN: 979-8-3597-2874-4

| 1  | How _ To Guides | 7 |
| 2  | How to Change Characters | 7 |
| 3  | How to Play as Nightwing | 9 |
| 4  | How to Play as Red Hood | 21 |
| 5  | How to Play Co-op | 32 |
| 6  | How to Interrogate | 34 |
| 7  | How to Unlock Knighthood | 36 |
| 8  | Walkthrough | 39 |
| 9  | 01: Batman's Last Case | 39 |
| 10 | 02: The Rabbit Hole | 68 |
| 11 | 2.1 – AKA Oswald Cobblepot | 68 |
| 12 | 2.2 – The Powers Club | 77 |
| 13 | 03: In the Shadows | 85 |
| 14 | 3.1 – The Key | 85 |
| 15 | 3.2 – Chelsea Tunnel | 88 |
| 16 | 04: The Masquerade | 98 |
| 17 | 05: The Court of Owls | 110 |
| 18 | 06: Jacob Kane | 126 |
| 19 | 6.1 – Court Judgement | 126 |
| 20 | 6.2 – The Voice of the Court | 129 |
| 21 | 07: The League of Shadows | 148 |
| 22 | 7.1 – Friends In Need | 148 |

23    7.2 – Talia al Ghul                              149

24    8.2 – The Lazarus Pit                            157

25    08: Head of the Demon                            161

26    8.1 – Dangerous Skies                            161

27    Suit Styles                                      161

28    Characters                                       171

29    Batman (Bruce Wayne)                             171

30    Robin (Tim Drake)                                172

31    Nightwing (Dick Grayson)                         174

32    Batgirl (Barbara Gordon)                         175

33    Red Hood (Jason Todd)                            176

34    Harley Quinn                                     178

35    Clayface                                         179

36    Mister Freeze                                    180

37    Court of Owls                                    181

38    The Penguin                                      183

39    Skill Trees                                      184

40    How to Unlock Skill Trees                        184

41    How Skill Trees Work                             184

42    What is The Knighthood Skill Tree                184

43    Red Hood Skill Trees                             185

44    Batgirl Skill Trees                              193

45    Robin Skill Trees                                 197

46    Nightwing Skill Trees                             203

47    What are Momentum Abilities                       209

48    How to Unlock Momentum Abilities                  209

49    Crafting Guide                                    212

50    How Crafting Works in Gotham Knights              212

51    What Can You Craft in Gotham Knights              212

52    Which Character Should You Pick                   215

53    Robin                                             215

54    Batgirl                                           216

55    Red Hood                                          217

56    Nightwing                                         218

Gotham Knights guide and walkthrough will help to lead you through every step of Gotham Knights' story whether you're playing solo or co-op, including solving detective sections, taking out Gotham's most wanted villains, and uncovering the secret plot of the Court of Owls.

This walkthrough will include the best ways to take down thugs and their bosses, and making sure you don't miss any secrets or collectibles along the way.

In this guide, you will primarily find information on how to complete the game - a full walkthrough. In addition, you will learn a bit about the playable heroes: Robin, Batgirl, Nightwing and Red Hood, as well as their skills and gadgets. In the Beginner's guide section, you will learn tricks and tips on how to get started and develope your heroes.

# HOW _ TO GUIDES

## How to Change Characters

This Gotham Knights guide will show you how to change characters and choose between Robin, Batgirl, Red Hood, and Nightwing. It's still possible to swap characters after your initial choice is made, so don't worry – you just won't be able to do it immediately.

You can change characters by going to the Belfry, which is the base of operations for the team. The first time you enter the Belfry will be at night when nearly everything inside is offline or unavailable to interact with.

The option to change characters will unlock after you've completed the first tutorial night patrol for whichever character you chose to use after the opening cutscenes. So, you're only stuck as your first choice for one night in Gotham City.

After that first night, you'll find yourself inside the Belfry during the daytime. Now you can head to the back-left corner of the room to find the hero suits for Batgirl, Robin, Red Hood, and Nightwing. To switch characters, simply interact with the suit that belongs to the character you want to switch to.

Gotham Knights features a system that has you patrolling the city streets on a nightly basis, so before each night's patrol, you're given the opportunity to switch characters in the Belfry.

When it comes to gaining experience and leveling up, there's no need to worry about switching often because each character will level up at the same time. This means that you'll just need to distribute any skill points that you've earned with one character into the others' skill trees once you've switched to them.

Thankfully, you aren't severely punished for switching characters from night to night. With that being said, sticking with one character for a while will help to unlock their unique Knighthood skill tree by completing specific challenges.

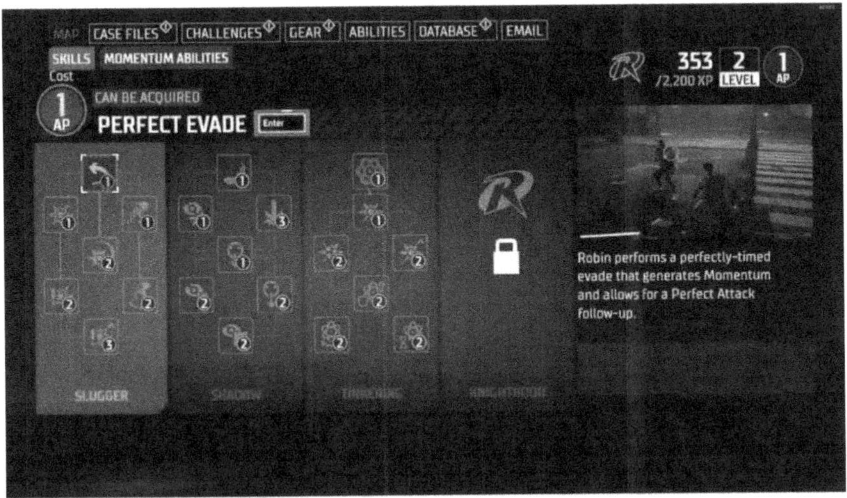


If you're unsure of which character to switch to, be sure to check out our guide on which character you should pick. It's useful if you're looking for a quick rundown of each character's abilities, strengths, and play styles, as the game doesn't exactly provide much information to work with during the initial choice.

## How to Play as Nightwing

Among Gotham Knights' four protagonists, Dick Grayson (Nightwing) is sure to win over fans of acrobatic combat with his exclusive skill trees.

After the death of Batman, the player must explore an open-world Gotham City as they fight crime and unlock the potential of each of the four protagonists: Nightwing, Red Hood, Batgirl, and Robin. Having four heroes to play in Gotham Knights means each player will be able to explore their different playstyles. The game's developers have said that this way of telling the story is meant to encourage exploration and experimentation. Each of the four protagonists brings something new to the table, and fans of acrobatic fighting styles will likely be drawn to Nightwing.

### Who Is Nightwing in Gotham Knights?

Dick Grayson, also known as Nightwing, was Batman's first protégé and the first one to don the "Robin" title. He grew up in a circus' acrobat family and is of the belief that fights are only worth it if there is something worth fighting for. After leaving Batman's tutelage, he became the hero and vigilante known as Nightwing. Nightwing is a master in acrobatics and wields dual escrima sticks.

## How to Upgrade Nightwing's Abilities

There are four skill trees per character in Gotham Knights. Abilities in the different skill trees are unlocked using Ability Points, which are earned as the player progresses through the game. Some abilities have requirements that must be met to unlock them. This means the player must unlock certain skills or achievements before progressing further in the skill tree.

### *Nightwing's Raptor Skills*

Nightwing's Raptor skill tree is themed around those abilities that pertain to dealing critical hits against single targets. There are seven abilities in the Raptor skill tree, each with its own requirements and costs.

- Perfect Evade: A perfect evade followed by a follow-up attack

    - Requirements: Unlock the Raptor skill tree

    - Cost: Unknown

- Critical Expertise: Critical damage is increased by 20%

    - Requirements: Unlock Perfect Evade

    - Cost: 1 Action Point

- Assassin's Mark: Damage to a marked target is increased by 10%

    - Requirements: Unlock Perfect Evade

    - Cost: 2 Action Points

- Trampoline: After using the Pounce ability, Nightwing performs a high jump on the enemy.

- Requirement: Unlock Perfect Evade

- Must be in an area in which high jumps can be made.

- Cost: 1 Action Point

- Precise Strikes: Chance of landing a critical hit increases by 10%

  - Requirement: Unlock Critical Expertise

  - Cost: 2 Action Points

- Aerial Bounce: Allows for Nightwing to perform up to three bounces off enemies, followed by an aerial attack

  - Requirement: Unlock Trampoline

  - Cost: 2 Action Points

- Critical Distance: Hitting an enemy at a distance increases critical chance and critical damage by 15%

  - Requirement: Unlock Aerial Bounce or Precise Strikes

  - Cost: 3 Action Points

### Nightwing's Acrobat Skills

The seven skills in Nightwing's Acrobat skill tree are granted momentum by evading enemies' attacks, which then increase the abilities' strength. This allows players to engage with Nightwing's circus background while attacking enemies.

- Aerial Damage Plus: The damage done by aerial attacks increases by 20%

    - Requirement: Unlock the Acrobat skill tree

    - Cost: 1 Action Point

- Extra Momentum Bar: Adds a second momentum bar to Nightwing's arsenal

    - Requirement: Unlock Aerial Damage Plus

    - Cost: 1 Action Point

- Momentum Gain Plus: Nightwing's momentum is increased by 15%

    - Requirement: Aerial Damage Plus

    - Cost: 1 Action Point

- Evade Chain: Performing quick backward jumps triggers a chain of perfect evasions

    - Requirement: Extra Momentum Bar or Momentum Gain Plus

    - Cost: 3 Action Points

- Haly's Favorite: A successful evade chain knocks down nearby enemies

    - Requirement: Evade Chain

- ■ Cost: 2 Action Points

- Evade Chain Momentum: A successful evade chain restores some momentum

  - ■ Requirement: Evade Chain

  - ■ Cost: 2 Actions Bars

- Mind and Body: Momentum abilities restore some health

  - ■ Requirement: Haly's Favorite or Evade Chain Momentum

  - ■ Cost: 2 Actions Bars

### Nightwing's Pack Leader Skills

The skills in Nightwing's Pack Leader skill tree cater to multiplayer, allowing players to buff or heal their teammates in battle.

- Family Ties: Nightwing's resistance is increased by 10%, and his teammates also receive boosts (Batgirl receives a 15% boost to her melee damage, Red Hood receives a 20% boost to his ranged damage, and Robin receives a 15% boost to his stealth damage)

  - ■ Requirement: Unlock the Pack Leader skill tree

- Cost: 1 Action Point

- Health Bolstered Defense: Nightwing receives a 5% defense bonus (which scales to 20%) when his health is above 70%

  - Requirement: Family Ties

  - Cost: 1 Action Point

- Momentum Regen: Nightwing's momentum regenerates over time, and the speed of this regeneration is increased when working with others

  - Requirement: Health Bolstered Defense

  - Cost: 2 Action Points

- Shared Skill: Passive skills that increase a player's stats are shared with allies at 50% value

  - Requirement: Health Bolstered Defense

  - Cost: 2 Action Points

- Elemental Smart Darts: Elemental effect is built up on enemies struck by Nightwing's darts

  - Requirement: Momentum Regen OR Shared Skill

  - Cost: 2 Action Points

- Elemental Smart Darts Plus: Darts reduce enemies' defense by 10% and increase damage done by allies by 5% for ten seconds

  - Requirement: Elemental Smart Darts

  - Cost: 2 Action Points

- Revive Darts: Darts revive allies (one per night)

- ■ Requirement: Elemental Smart Darts

- ■ Cost: 2 Action Points

### *Nightwing's Knighthood Skills*

The Knighthood skill tree represents a moment of change for each of the four protagonists, which happens when they hit a key point in their individual character arcs. When they have these realizations, new skills are unlocked and added to the skill tree. Each protagonist needs to be pitted against several unique challenges in order to reach Knighthood.

- Flying Trapeze: Nightwing uses his Flying Trapeze, a call back to his days in the circus, to glide from building to building

  - ■ Requirement: Unlock the Knighthood skill tree

- Triple Darts: 3 darts can be shot instead of 1

  - ■ Requirement: Flying Trapeze

  - ■ Cost: 1 Ability Point

- Strike Distance +: Nightwing can strike with melee attacks from farther away

- Requirement: Flying Trapeze

- Cost: 1 Ability Point

- Guardian: Nest Ability cooldown is reduced by 15% upon defeating an enemy

  - Requirement: Flying Trapeze

  - Cost: 1 Ability Point

- Bigger Nest: Increases Nightwing's nest size

  - Requirement: Guardian

  - Cost: 3 Ability Points

- Nest Buffs +: Nest damage is increased by 150%, and the healing effect is increased by 100%

  - Requirement: Strike Distance +

  - Cost: 3 Ability Points

- Combat Expertise: Increases amount of attacks in Nightwing's melee combo by one, and the last hit is a knockdown

  - Requirement: Triple Darts

  - Cost: 2 Ability Points

## Nightwing's Momentum Abilities

Nightwing's momentum abilities are unlocked by completing challenges unique to his character.

- Shotgun Darts: Nightwing shoots a dart from each hand, which do greater damage to closer enemies

  - Momentum cost: 1

- Elemental Shockwave: Smashing the ground creates an elemental trail that damages enemies

  - Momentum cost: 1

- Pounce: Nightwing jumps at a ranged enemy, inflicting damage and knocking them down

  - Momentum cost: 1

- Nest: Nightwing creates a nest around himself and his allies, granting a defensive bonus and healing over time. The nest damages any enemies that breach it and induces fear.

  - Cooldown: Five minutes

■ Momentum cost: N/A

## How to Play as Red Hood

With Batman gone, Red Hood is now one of the Gotham Knights ready to protect Gotham. Here is how to play Red Hood and become the next Dark Knight.

Gotham Knights is almost here, and fans of the Batman: Arkham series are excited to see what the game offers. While the game does not occur in the Arkham verse, it does take place after Batman's death. It follows Nightwing, Robin, Batgirl, and Red Hood, as it is up to them to protect Gotham City and ultimately decide who is next to take up the Batman mantle.

A surprising addition to the team is Red Hood. For years Red Hood has been more of an anti-hero using lethal force to take down his enemies and go against Batman's ideals, all because of his dark past involving Batman. This has turned Red Hood into a cult-following hero, and fans are excited to see how he plays Gotham Knights, as his fighting style involves a lot more range combat and brute strength. Not all information is currently available, so this article will be updated regularly.

## Who Is Red Hood in Gotham Knights?

Red Hood, also known as Jason Todd, was the second person to take up the Robin mantle after Dick Grayson became Nightwing. Jason was far more hot-headed than Dick and seemed to get into more trouble as Robin. Batman believed he could correct his behavior and save the boy from his own eventual destruction. Unfortunately, Batman failed when Jason confronted the Joker alone and died in a fiery explosion along with his mother.

Six months later, Jason was resurrected due to the events involving Superboy-Prime shattering reality. Talia al Ghul discovered him and used the Lazarus Pit to restore his memories. At this point, Jason hated Batman and his ideals, believing the only way to stop villains is to kill them for good. In Gotham Knights, Batman's death seems to have struck a chord with Red Hood, as he is on a path toward redemption and trying to become the next protector that Gotham City needs.

## *How to Upgrade Red Hood's Abilities*

In Gotham Knights, similar to other RPGs, players will level up their characters the more they play them. With each level up, the player will be granted ability points toward each character's skill trees that unlock new moves and abilities to use during combat and explore all of Gotham. Each character has three skill trees that can be upgraded, all with their own unique abilities depending on the character. In Red Hood's case, he has a marksman tree, a brawler tree, and a vengeance tree. Additionally, each character has momentum abilities and knighthood skills.

### *Red Hood's Marksman Skills*

Marksman skills not only improve Red Hood's damage but also his skills with his firearms.

- Perfect Evade: A perfect evade followed by a follow-up attack

    - Requirements: Unlock the Marksman skill tree

    - Cost: Unknown

- Focused Fire: Aim longer at a target to inflict 4x damage

    - Requirements: Perfect Evade

    - Cost: 1 Action point

- Lucky Rounds: A slight chance to fire rounds that do five times the damage when applied to Ranged Attacks or Precision Aim

    - Requirements: Perfect Evade

    - Cost: 2 Action points

- Critical Expertise: Increases critical damage by 20%

    - Requirements: Perfect Evade

- Cost: 1 Action point

- Precise Strikes: Increases chance of landing critical hits by 10%

    - Requirements: Critical Expertise

    - Cost: 2 Action points

- Focused Fire +: Reduces the amount of aim time for Focused Fire by 50%

    - Requirements: Focused Fire

    - Cost: 2 Action points

- Quickfire Expert: Increases the critical chance and critical damage of Ranged Attacks by 15 percent

    - Requirements: Precise Strikes or Focused Fire +

    - Cost: 3 Action points

### Red Hood's Brawler Skills

Red Hood's Mystical Leap in Gotham Knights makes him even more badass

Red Hood is the biggest-looking character compared to his allies, meaning he comes with some brawling skills to go along with his firearms.

- Human Bomb: When throwing an enemy, a concussion mine is attached that explodes when shot

    - Requirements: Unlock the Brawler skill tree

    - Cost: 3 Action points

- Large Grab: Perform grab moves on large enemies

    - Requirements: Human Bomb

    - Cost: 1 Action point

- Extended Grab Window: Grab an enemy at 50% or less health

    - Requirements: Human Bomb

    - Cost: 1 Action point

- Human Bomb Enhanced: Increases the damage and radius of the Human Bomb

    - Requirements: Large Grab or Extended Window Grab

    - Cost: 2 Action points

- Grip Expertise: Increases grab move damage by 10%

    - Requirements: Human Bomb Enhanced

    - Cost: 1 Action point

- Iron Grip: Prevents grab moves from being interrupted by most attacks

    - Requirements: Human Bomb Enhanced

- ■ Cost: 2 Action points

- Human Bomb Multiplied: Once a concussion mine is detonated, six additional concussion proximity mines are deployed on the ground

    - ■ Requirements: Iron Grip or Iron Grip Expertise

    - ■ Cost: 2 Action points

### Red Hood's Vengeance Skills

Red Hood's Vengeance tree focuses more on Jason's need for revenge against those who have wronged him and his family. This focuses on damage to certain enemy types.

- Coup de Grace: Inflicts 10% more damage to enemies with 30% health or less

    - ■ Requirements: Unlock the Vengeance skill tree

    - ■ Cost: 1 Action point

- Freak Justice: Increases damage by 15% and critical damage by 5% against Freaks

- ■ Requirements: Coup de Grace

- ■ Cost: 1 Action point

- Mob Justice: Increases damage by 15% and critical damage by 5% against the Mob

  - ■ Requirements: Freak Justice

  - ■ Cost: 2 Action points

- Regulator Justice: Increases damage by 15% and critical damage by 5% against Regulators

  - ■ Requirements: Freak Justice

  - ■ Cost: 2 Action points

- Combined Fire: Increases damage of Ranged Attack combos and Precision Aim for Red Hood and his allies when focused on a single enemy

  - ■ Requirements: Mob Justice or Regulator Justice

  - ■ Cost: 2 Action points

- Unrestricted Fire: Shoot unlimited rounds for a short while after performing a Two-Fisted Reload

  - ■ Requirements: Combined Fire

  - ■ Cost: 2 points

- Double Vortex: Shoot double the number of rounds after performing a Two-Fisted Reload

  - ■ Requirements: Combined Fire

  - ■ Cost: 2 points

## *Red Hood's Knighthood Skills*

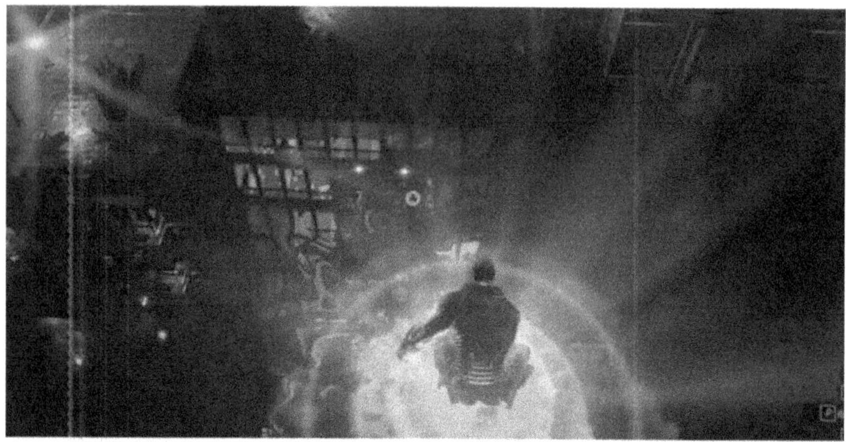

Red Hood's Mystical Leap in Gotham Knights makes him even more badass

Knighthood skills only unlock after progressing through the story, once the character discovers they are meant to become the next Dark Knight.

- Mystical Leap: Travel through the air using spirit platform

    - Cost: Unknown

- Ranged Terror: Shots that knock out targets and inflict fear on nearby enemies

    - Cost: 1 Action point

- Weak Spot Damage +: Increases damage by 15% for headshots and weak spots

    - Cost: 1 Action point

- Grab Dread: Inflicts Fear in nearby enemies when grabbing a target

- ■ Cost: 1 Action point

- Combat Mastery: Increases the number of attacks in the ranged attack combo by 1

  - ■ Cost: Unknown

- Duncra's Training: Mystical rounds require less time to lock onto targets by 50 percent

  - ■ Cost: Unknown

- Shadow Vengeance: Increases Mystical Rounds shots from 1 to 2

  - ■ Cost: 3 Action points

### Red Hood's Momentum Abilities

Momentum abilities are a unique set of moves for specific characters depending on their specialization. For Red Hood, that's firearms and brute strength.

- Barrage: Shoot multiple rounds straight ahead, hitting everything in their path, and then reload

- Cost: 1 Action point

- Mystical Punch: Charge up mystical damage to his fists for a 50% boost to melee damage and 5x more Elemental Buildup for 30 seconds

  - Cost: 1 Action point

- Two-Fisted Reload: Quickly reload while hitting and shooting at an enemy

  - Cost: 1 Action point

- Harnessed Rage: Next attack increases damage by a portion of damage received while healing 1% per second for 20 seconds

  - Cost: 1 Action point

- Spoilsport Reload: Jump backwards to release explosive magazines and reload

  - Cost: 2 Action points

- Portable Turret: Deploys high-damage mini-turret to shoot at enemies for 6 seconds

  - Cost: 2 Action points

- Mega Tackle: Dash at enemies in his path and knock them down

  - Cost: 2 Action points

- Mystical Rounds: Lock-on ability that shoots massive-damage mystical rounds at all targets

  - Cost: Unknown

# How to Play Co-op

Gotham Knights allows you to play in co-op with friends, but multiplayer isn't an option that's available from the get-go. There's a short tutorial mission that needs to be completed before you can freely hop online, but the game will take a moment to let you know when the multiplayer options have been unlocked.

Co-op multiplayer will unlock after your first night on patrol in Gotham City. Before you can invite friends or join online sessions, you'll need to go through that first tutorial night for whichever character you chose first after the opening cutscenes.

When you return to the Belfry – the Batfamily's main base of operations – it'll be daytime. Complete the mandatory tasks assigned to you in the Belfry and you'll be able to return to the streets of Gotham City at night in co-op multiplayer.

Press Left on the D-Pad with a controller or C on PC to open up the Social Wheel to quickly adjust online privacy settings after unlocking multiplayer. At this point, you can also bring up the pause menu to open up the full Social and Multiplayer menus. The Social Wheel also contains shortcuts to get to the game's photo mode and emote menu.

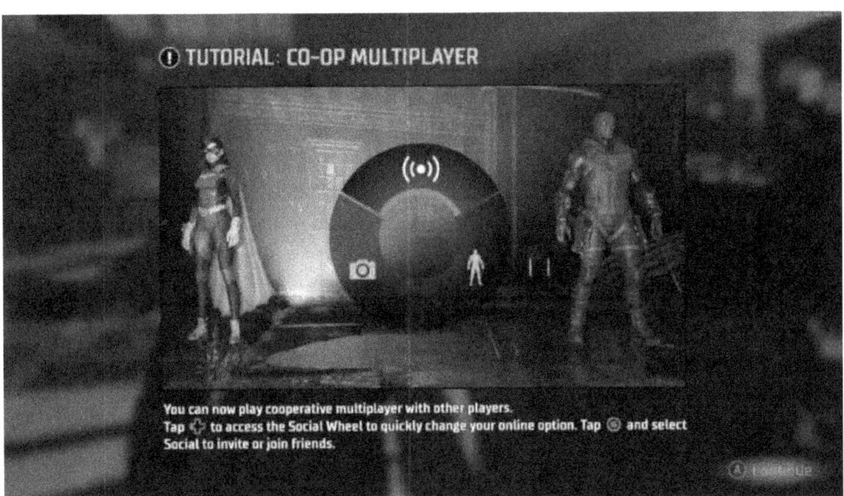

The full Social menu will allow you to invite friends, join their games, block other players, and more. Meanwhile, the Multiplayer menu provides the option to drop in on random online sessions through matchmaking.

Both of these menus will be locked until you reach the point in the story where the Belfry is accessible during the day, but this shouldn't take more than an hour or so to reach unless you decide to explore Gotham City during that first night.

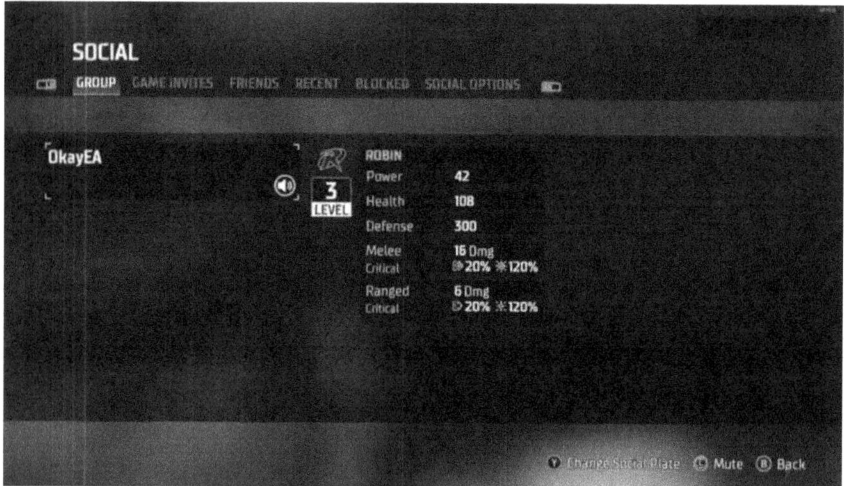


Once you finish up that tutorial night, you'll also be able to change your character and swap to any of the other Knights. Many more gameplay and customization options become available if you progress through those first few tasks at the beginning of the game.

If you're having trouble deciding which team member to settle on, be sure to check out our guide on which character you should pick for some useful advice.

# How to Interrogate

This Gotham Knights guide will show you how to interrogate enemies on the streets of Gotham City. As a member of Batman's found family, it's important to know how to ask a criminal for information while they're on the verge of passing out from severe pain.

You'll learn how to interrogate enemies after you've completed your first night on patrol in Gotham City as the character you chose at the beginning of the game. Learning how to perform interrogations is a mandatory quest objective, but we're here to help just in case you still have some questions.

First of all, you can't just interrogate any enemy who shows up for a fight. Hold X on PC or Down on the D-Pad with a controller to scan criminals and figure out which one is an informant. Informants are marked by specific icons that look like an upside-down triangle with a question mark inside.

Once you identify the informant, either approach using stealth or lower their health in combat. Then, when you get close enough, you can grab the informant by pressing R2 on PlayStation, RT on Xbox, or by holding down Tab on PC. Then, press Triangle, Y, or the Q key to               interrogate               the               informant.

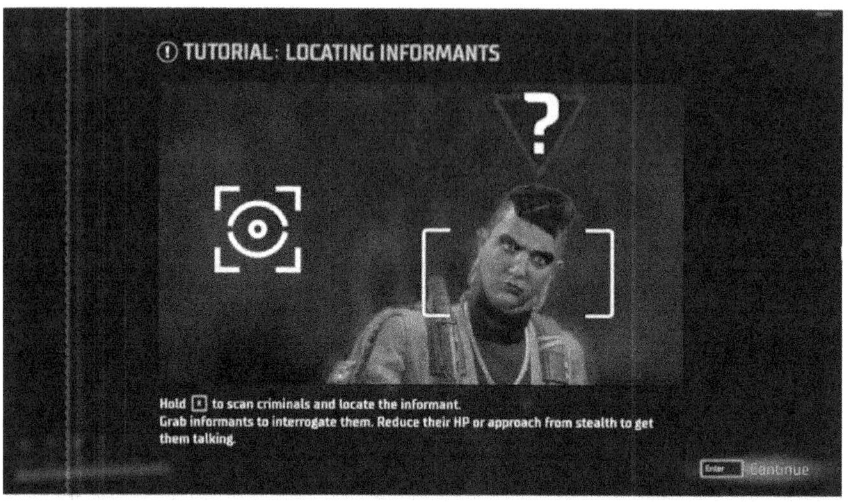


While stealth is usually the best option if your main goal is to interrogate, you can also grab criminals who are stunned, low on health, or grounded by certain takedowns. It's unlikely that you'll be able to stealthily grab an informant who's surrounded by a group, so, sometimes, open combat is the only option. Just be sure to keep an eye on the target so that they don't get knocked out by your combos.

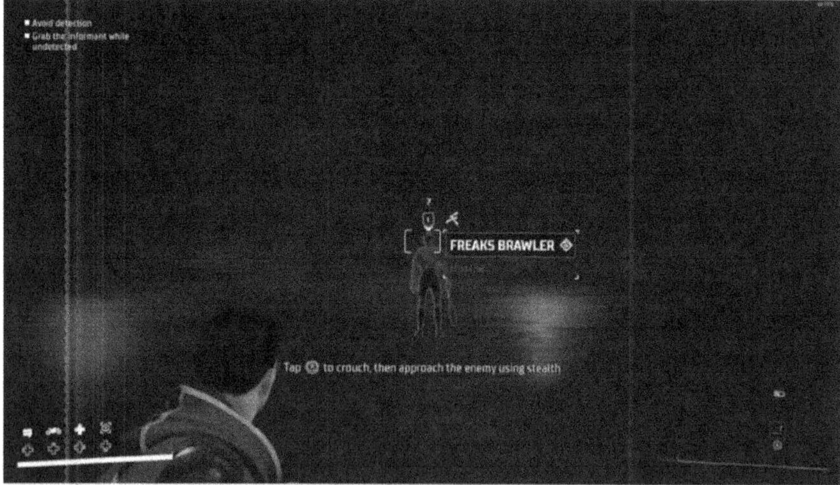


It's much easier to land stealth interrogations with Robin and Batgirl, as they both have abilities in their skill trees that better facilitate stealth gameplay and tackling challenges from the shadows. Nightwing and Red Hood are more suited to the loud approach, but the job can get done either way.

If you're still unsure of which Knight's play style suits you best, check out our guide on which character you should pick for some helpful advice.

## How to Unlock Knighthood

This guide will explain how to unlock the Knighthood skill tree in Gotham Knights. Each playable character has access to three skill trees when the game begins, and Knighthood is the fourth, locked one. It contains powerful and useful abilities that you'll want to gain access to as soon as possible.

To unlock the Knighthood skill tree, you'll need to complete Knighthood Challenges. These challenges will only appear after you've gone through the first night on patrol in Gotham City as the first character you chose at the beginning of the game. That short tutorial patrol should only take around an hour or less to complete.

At the end of the night, you'll return to the Belfry – the Batfamily's base of operations. This time, you'll be inside the Belfry during the day. Go through all the new, mandatory tasks in the Belfry to gain access to the Knighthood Challenges in the Batcomputer.

Go to the Challenges tab in the Batcomputer to find the list of Knighthood Challenges. These need to be completed for each character individually to unlock their respective Knighthood skill trees.

Using Robin as an example, his Knighthood Challenges require him to complete Timed Strike Training in the Belfry, stop 10 premeditated crimes, and defeat three large, shield-wielding enemy types. After these tasks are taken care of, his Knighthood skill tree will be unlocked.

Unlocking Knighthood not only provides new skills to use, but also grants access to Heroic Travel options that allow you to traverse Gotham City much faster than you normally would.

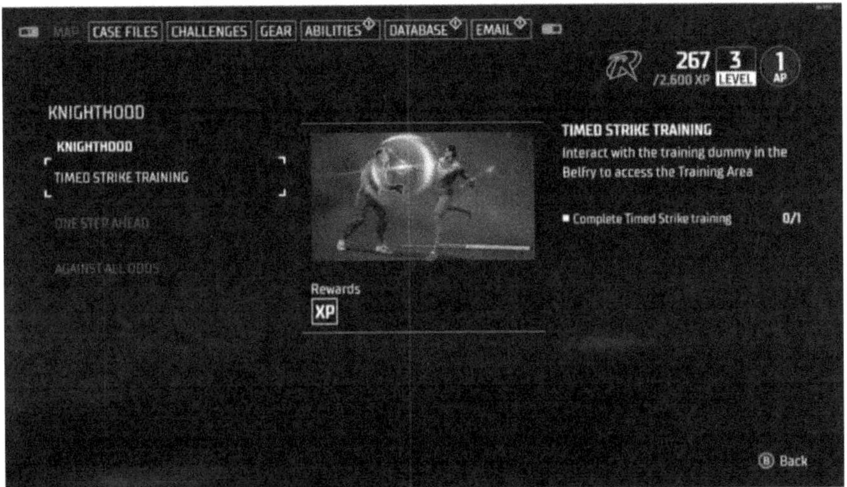


Once you arrive at the Belfry during the day, you also gain the option to change your character before going back out on patrol at night in Gotham. You aren't punished for regularly swapping characters, but we recommend that you stick with Robin, Batgirl, Nightwing, or Red Hood long enough to unlock Knighthood for each of them since the skill tree is so useful.

If you're unsure of which Knight to stick with first, you can have a look at our guide on which character you should pick to get a brief description of their strengths and abilities.

# WALKTHROUGH

## 01: Batman's Last Case

### 1.1 - Kirk Langstrom

Kirk Langstrom is the first mission in Gotham Knights' opening Case File, Batman's Last Case, and it'll serve as a tutorial that'll teach you many of the core gameplay mechanics you'll need to master to keep the citizens of Gotham safe.

In this walkthrough, you'll be provided detailed instructions on how to complete Gotham Knights' opening mission, Kirk Langstrom.

*Spoiler Warning!*

*This walkthrough contains Gotham Knights' spoilers. Read with caution if you've yet to roll credits.*

*Kirk Langstrom Walkthrough*

Prior to his death, Batman initiated Code Black, a fail-safe that destroyed the Batcave and simultaneously transmitted a video message to his apprentices. Attached to the transmission was an encrypted case file that mentions Dr. Langstrom, a zoologist at Gotham University.

To start your journey and investigate the mysterious Dr. Langstrom, select one of the four playable characters: Batgirl (Barbara Gordon), Nightwing (Dick Grayson), Red Hood (Jason Todd), or Robin (Tim Drake).

If you're unsure who to pick, check out our Which Character Should You Pick guide to learn more about the young heroes including their strengths and playstyles.

Though you'll be able to freely change your character later, the character you select here will be locked-in throughout the entirety of mission 1.1 - Kirk Langstrom.

With your character selected, you'll now learn Gotham Knights' base movement controls including how to mantle, grapple, perch, and jump across large gaps.

After traversing the opening area and learning of Dr.Langstrom's death, head toward his office by grappling onto the nearby American flag and then onto the Moulton Hall Balcony. If you need assistance locating the balcony or the American flag, search for the small diamond-shaped icon pictured above and it'll guide you in the right direction.

Search the hall for Langstrom's office and once inside, use your AR Targeting to scan the area for clues. By scanning the floor near Langstrom's desk, you'll discover a yellow trail of fresh scratch marks. Follow the trail and you'll be led to the crime scene where Langstrom died.

Approach the bloody crime scene to further investigate, then use your AR Targeting to mark a hidden mechanism beneath Langstrom's cabinet. To analyze the mechanism, hold your AR Targeting and aim your reticle at the object.

Next, approach the nearby desk to investigate the lab's workstation.

*Investigate the Crime Scene*

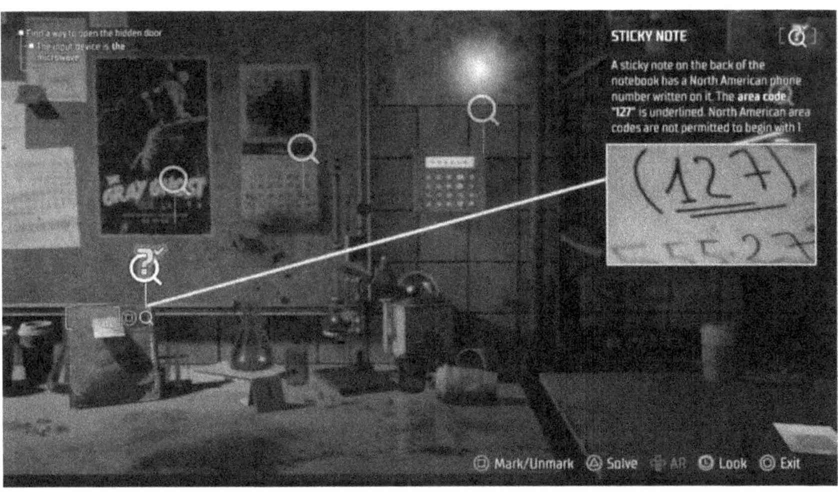


Here you'll learn how to investigate crime scenes – a puzzle mechanic that tasks you with exploring an area to solve a pertinent question, in this case, how to open Langstrom's hidden door. The questions seen on the left side of your screen will serve as clues that can help you narrow down what to search for.

The questions for this crime scene read:

- Which input device unlocks the hidden door?

- What is Langstrom's secret passcode?

This indicates that you'll be looking for a numerical passcode and a device in which to input said passcode. To explore the area for clues, move your reticle around and use AR to highlight different objects to learn more about each one and how they may help solve this puzzle. If you think you know which two objects in the room can combine to reveal the solution, mark the two objects then press "Solve".

There is no penalty for incorrect answers and you'll be able to freely select new object combinations if your initial solution was wrong. If you're simply looking for the solution reveal the spoiler below.

*HIDE SPOILERS*

*Mark the yellow sticky note with the number "127" written on it, and mark the microwave before pressing "Solve".*

Walk through the newly revealed passage and explore the area before picking up the hard drive sitting on Langstrom's workstation.

With the hard drive secured, retrace your steps to leave the Gotham City campus. On your way out you'll be confronted by a Brawler — a grunt from the Freaks faction that only utilizes melee attacks. Use your Melee, Heavy Melee, and Evade moves to take them down.

Next, enter Hummel Hall to learn how to crouch, how to perform Ambush Takedowns, and how to perform Silent Takedowns. With the Freaks in Hummel Hall defeated, exit the building by grappling out of the window and onto the statue in the courtyard. Tap your AR to ping all of the enemies in the area and take them out before entering Woolfolk Hall.

Inside you'll learn how to use Ranged Attacks and you'll be introduced to Firestarters – members of the Freaks faction that primarily use ranged attacks to throw Molotovs that can damage anyone within the attack radius.

Take down the horde of enemies in the library and rescue the hostages to trigger the entrance of a Bulldozer Freak. To inflict damage on shielded enemies like Bulldozers you'll first need to use Heavy Melee Attacks to break their guard and they'll then become vulnerable to subsequent strikes.

Defeat the Bulldozer while remaining cautious of their red-color-coded attacks. In Gotham Knights, all attacks that appear in red cannot be blocked or interrupted but can be evaded.

After defeating the Bulldozer, exit the building, summon the Batcycle, and follow the GPS Navigation to reach The Belfry's east plaza entrance. As briefly mentioned in Batman's farewell transmission, The Belfry will serve as your base of operations throughout your campaign.

Once inside The Belfry, you'll view a brief cutscene. Congrats, you've now completed mission Kirk Langstrom in Gotham Knights' opening chapter, Batman's Last Case!

## 1.2 - The Langstrom Drive

The Langstrom Drive is the second mission within Gotham Knights' opening Case File, Batman's Last Case, and it sees the squad further investigating the enigmatic Dr. Langstrom.

In this Gotham Knights walkthrough, you'll learn how to complete The Langstrom Drive's puzzles and detail how to punch your way through the various enemies that'll stand in your way.

*Spoiler Warning!*

*This walkthrough contains Gotham Knights' spoilers. Read with caution if you've yet to roll credits.*

*The Langstrom Drive Walkthrough*

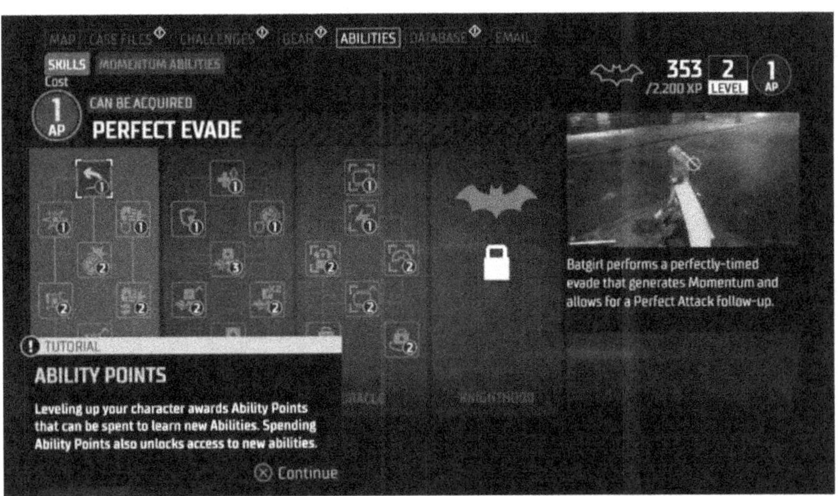

The Langstrom Drive opens with the gang convened inside The Belfry after recovering Dr. Kirk Langstrom's encrypted hard drive. Begin this mission by interacting with the Batcomputer so you can spend Ability Points in the Abilities menu to unlock Perfect Evade. When your evade is perfectly timed, you'll earn bonus Momentum to build toward your Momentum Abilities and you can perform a Perfect Attack follow-up.

You'll subsequently unlock your first Momentum Ability. Momentum Abilities are powerful, character-specific moves that can only be used after accruing sufficient Momentum which is earned by landing attacks.

If you purchased the Gotham Knights Deluxe Edition you can now equip your Beyond Suitstyle, if desired.

After giving all of your menus a sufficient look-through, exit The Belfry and start taking down criminals to gather Clues while the rest of the gang works to get The Belfry back online.

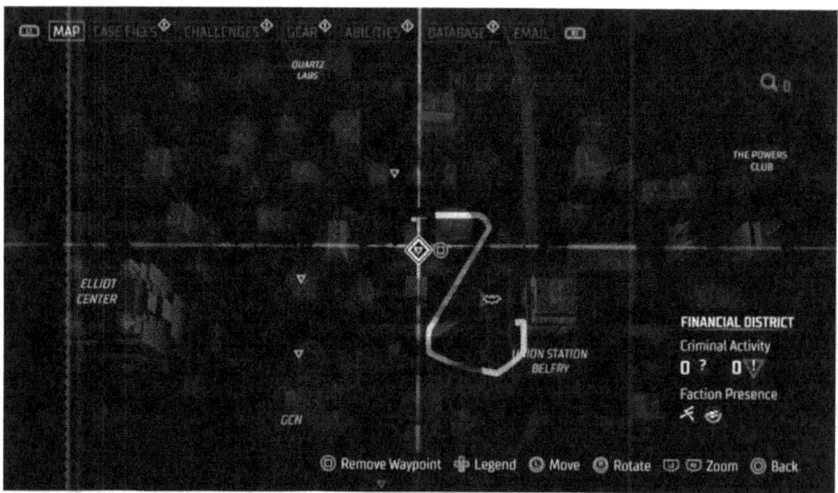


To identify criminal behavior you can either check your Map for crime markers or tap your AR scan to display nearby Opportunistic

Crimes on your compass.

As you defeat more criminals some of them will drop Clues, that when collected will reveal Premeditated Crimes that you can thwart during your next patrol. Once you've defeated enough criminals, Alfred will radio in with a message for you.

Once you've listened to the message, head toward GCPD which will already be marked on your compass and Batcycle GPS.

Upon your arrival, use your grapple to climb the GCPD rooftops while heading toward the diamond-shaped marker on your screen. The diamond marker will lead you directly to the back entrance of GCPD Headquarters.

The back entrance is a garage area filled with cops that wouldn't be happy to see a vigilante in their midst. Move through the area crouched and use your AR scan to track their movement. When an opening arises, head toward the door icon seen in the image below.

While you could brawl your way through the police officers, remember that you do not gain XP when battling the GCPD.

Interact with the marked door to enter GCPD Headquarters and as you move forward you'll find that the door to the security station is locked. To get around the door, look to the left and grapple into the conveniently open vent.

After crawling through the vents you'll find yourself in front of a few interrogation rooms. Proceed through the Holding Cells doors to enter the cell block filled with prisoners.

To sneak past the inmates, slip to the far left of the cells while remaining vigilant for cops on patrol then grapple up to the green-lit second story.

Your next objective and a Chest can be found to your right, but before proceeding to the path, you may want to investigate some of the items in the office to your left.

*Reaching the Morgue*

To reach the Morgue in the Forensics Wing, you'll need to clear the officer bullpen which is chock full of cops and security cameras. While there are tons of ways to reach the Forensics Wing, one method to avoid detection is to mantle through the left side of the room before grappling into the window pictured above.

Here you'll learn that the door to the Forensics Wing is locked so you'll have to collect the desk sergeant's keycard. Walk down the stairs and approach the Camera Control Panel below to disable all of the cameras in the room. With the cameras down you'll have some more leeway to snag the keycard from the sergeant's desk which is located in the center of the bullpen.

Sneakily return to the locked door and walk through the Forensics Wing to reach the Morgue where you'll find Talia al Ghul. After a brief conversation with Talia, use your AR to locate Langstrom's body and approach the autopsy table to search for Langstrom's Biodecryption Key.

*Finding the Biodecryption Key*

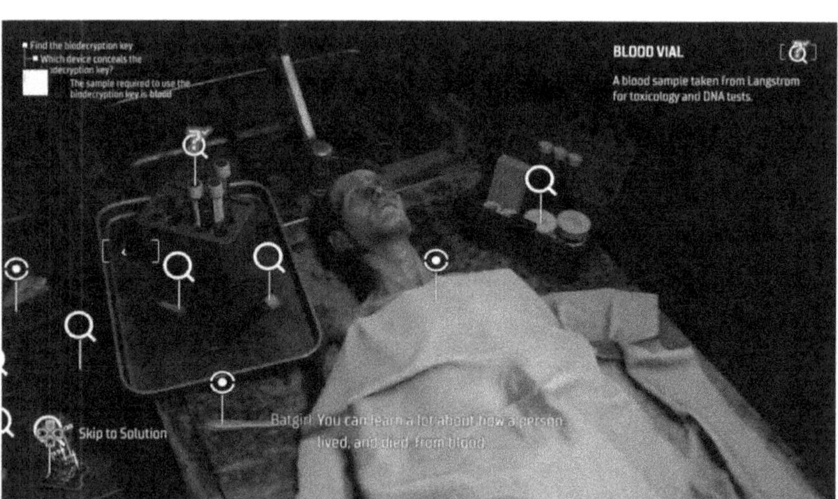


Before investigating the scene, consider the questions on the left side of your screen that read as follows:

*Which device conceals the biodecryption key?*

Which sample is required to use the biodecryption key?

With these questions in mind, you should be searching for a device powered by genetic material and the device's corresponding bio-material. Use your reticle and AR to investigate. Hovering over any of the clues and items in the area will provide you additional info, in the form of voice lines and text, so you can better determine whether they're relevant to the biodecryption key.

When you feel you've found the answers to the questions above, mark the two items and select, "Solve". If you'd like to bypass this puzzle and you're looking for the solution, reveal the spoiler hidden below.

*SHOW SPOILERS*

*Mark both the green-capped Blood Vial and the Blood Glucometer before pressing "Solve".*

After solving the puzzle, alarms will blare in GCPD Headquarters. Leave the autopsy room and sneak past the officers by using your grapple to stay out of their line of sight.

Your goal is to reach the exit just below the grapple point circled in the image above. Thankfully, your grapple has a rather generous range so you'll be able to safely grapple out of GCPD so long as no cops are stationed below the grapple point.

Now that you've exited the bullpen you're safe to leave the building as no enemies can be found in the area ahead. Zoom past the GCPD officers on your Batcycle and follow the GPS to make your way back to The Belfry.

This concludes mission 1.2 - The Langstrom Drive in Gotham Knights' first Case File, Batman's Last Case!

## *1.3 - Weird Science*

*Check the Evidence Boards*

From inside the Belfry, approach the Evidence Boards left of the console.

Highlight Clues in the Leads Section to Learn More About your Next Steps and Open the Challenge Menu

The Evidence Boards track information discovered during your investigation. The Leads section features the most relevant information about your next steps. Highlight evidence to gain information about the next steps of your investigation. Hover over the Leads section and press ⓪ / ⓦ . For the next step of your mission, we need to find criminal informants from the Freaks and the Mob and interrogate them to find out more information, in Robinson Park and West End respectively.

Prepare for Patrol – Craft a New melee Weapon Using the Workbench – Complete Interrogation Training – Read the Newspaper on the Workbench

Once you've examined the Leads section in the Evidence Boards, you need to get ready for patrol by carrying out three tasks. First, approach the coat hangers behind the Evidence Boards to enter the Training Area, then Basic Training, then Interrogation.

During the Training, we will learn how to locate criminals and the informant hiding in their group. To interrogate an informant, you need to grab them, but first you need to reduce their HP or stealthily approach them to get them talking.

Identify the Informant – Avoid Detection

During the training, follow the on-screen prompts, so hold ✥ for AR Targeting and then the right stick to aim to scan enemies and identify the informant.

*Avoid Detection – Grab the Informant While Undetected*

Next, R3 to crouch and approach the informant using stealth. When you're close enough to him, press R2 / RT to grab him.

*Avoid Detection – Interrogate the Informant*

Now △ to interrogate.

*Lower the Enemy's Health*

You can also grab enemies who are unaware, stunned, low on health, or affected by certain knockdowns. Approach the enemy and start weakening him until the R2 / RT prompt to grab him appears.

*Grab the Informant – Interrogate the Informant*

Grab the informant and then ⃝ / ⃝ to interrogate him.

Prepare for Patrol – Craft a New melee Weapon Using the Workbench – Read the Newspaper on the Workbench

Now approach the workbench right of the console and read the newspaper. This will trigger a new cutscene.

Prepare for Patrol – Craft a New melee Weapon Using the Workbench

Finally use the Workbench next to the newspaper to start crafting. In the Crafting menu, select Melee Weapon and press ⊗ / Ⓐ . Now hold ⊗ / Ⓐ to craft the Melee Weapon.

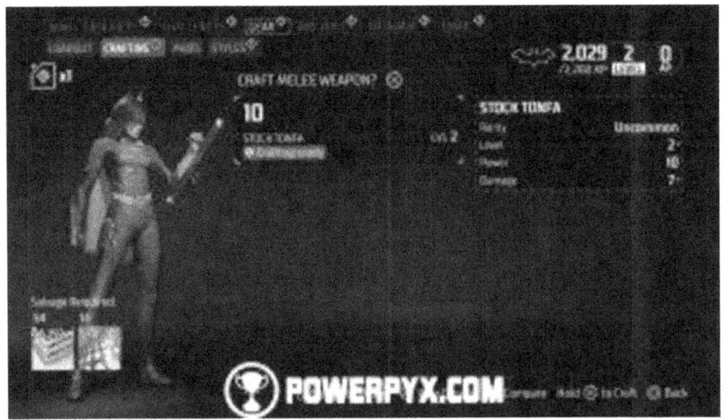

*Exit the Belfry and Patrol Gotham City*

Time to leave the Belfry and patrol the city for more clues. Head to the West End location in the first map screenshot below where there should a Mob crime going on. Weaken the enemies and interrogate one to complete the Snitches challenge.

Then, head to the Robinson Park location in the second map screenshot below and repeat the same to complete the A Little Chat challenge.

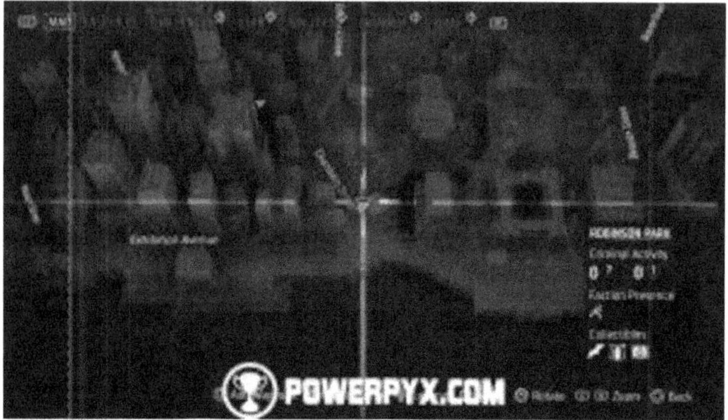


Now we have enough information to continue on.

*Talk to Talia al Ghul*

Proceed to Talia's location in Lower Gotham.

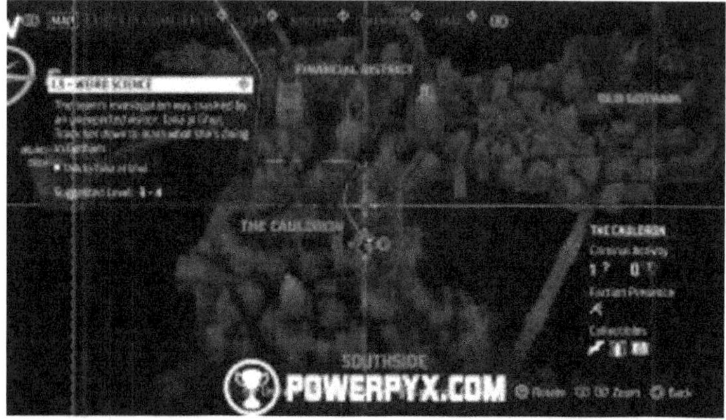


*Return to the Belfry*

Now we're ready to get back to the Belfry to regroup and fill the others in on the recent findings. Open your map and 🄡🄣 to fast travel to it. It also appears that Harley Quinn had made contact with the team.

## 1.4 – Blackgate Blues

Reach Blackgate Penitentiary

Exit the Belfry and make it to Blackgate Penitentiary where Harley Quinn is being held prisoner.

When you reach the main gate, crouch to avoid getting spotted by the patrol outside and continue down the stairs to the left to get under the bridge.

Continue under the bridge to get to the other side and then shimmy along the wall. Continue along the coast until you can enter the sewers.

*Find a Way into the Prison*

Continue on, squeeze through, climb up the ledge and you'll find a chest.

Then, grapple to the other side and continue on until you can interact with a panel. This will get the door unstuck and allow you to drop down.

*Reach Cell Block 3*

Crouch under the opening and you'll be in a guarded room. Since the cell block access is on the ground level, drop down and get rid of all guards. Once you have, get to the back of the room to find a chest. Another chest is at the end of the hallway left of the entrance to Cell Block 3.

*Now proceed to Cell Block 3*

Locate Harley Quinn's Cell – Clear the Inmates from the Corridor

After opening the doors, you'll find three inmates beating down a guard. Dispose of them and once you have, continue on to the elevator switch.

Defeat the Inmates – BONUS: Avoid Taking Damage

Drop down below and defeat the inmates. If possible, avoid taking damage for some extra XP.

Reach Harley Quinn's Cell

Now continue up the stairs.

Talk to Harley Quinn

Approach Harley's cell to talk to her.

Exit Cell Block 3

After talking to Harley, we're ready to leave the Cell Block.

Exit Cell Block 3 – Clear the Inmates from the Corridor

As you do, clear out the inmates from the corridors.

Exit Cell Block 3

Now squeeze through the door.

Cross the Yard – Defeat the Enemies

Clear out the prison yard of all enemies in order to proceed.

Cross the Yard

Cross the basketball court, making sure you first open the chest behind the inmates before continuing.

Reach the Basement

Continue along the corridors until a new cutscene starts during which a big Mobster woman shows up.

Defeat the Godmother

Get rid of the Godmother enemy any way you see fit. If she performs an Armored Attack, this can only be countered with Piercing Abilities, such as Batarang Barrage. Gain Momentum by landing melee attacks, then hold R1 / RB and tap ▲ / Y to unleash the Batarang Barrage.

Defeat the Guards 0/2

Two more guards will come investigating after you've defeated the Godmother. Defeat these as well.

*Reach the Basement*

Continue on by interacting with the elevator switch to open a door. More enemies in front of us to defeat. Next, continue up and down the elevator shaft, through and down the air vent.

Locate the Records Room – Defeat the Inmates 0/3

We're now in the basement, but there are three more inmates ambushing us. Defeat them to continue and remember to hold ⬛ / ✖ to perform a heavy melee attack against the Bulldozer enemy.

*Locate the Records Room*

Open the chest in the same room and continue on through the door until you reach the Records Room.

Scan the Card Catalog – BONUS: Find and Examine Marked Items 0/4

Before ✛ scanning the card catalog on the main desk, scan Old Newspaper (1/4) behind the main desk. Then, enter the top right room and scan Note from Warden (2/4) and Business Card (3/4). Next, enter the top central room to find a chest and Prisoner Profile (4/4).

Now scan the red-glowing items on the main desk.

Pick Up Card

Once scanned, you can pick up the card.

*Locate the Record*

Next, enter the room left of the top right room where we've scanned the 2 of the 4 items and scan the cabinet.

Pick Up File

Once scanned, pick up the file Harley was looking for from the drawer.

Reach the Nexus

Retrace your steps, through and up the air vent and up the elevator shaft.

Return the Document to Harley Quinn

Once you're back in the Nexus, more inmates will be fighting guards. Eliminate them all as you please and proceed through the prison corridors to get back to Harley.

Chase the Balloon

Harley has strung the book we need to a balloon that we need to chase. Run through the corridors, ignoring enemy fights.

Defeat the Guards and Rioting Inmates 0/12 – BONUS: Perform Perfect Evades 0/6

When you reach the courtyard, you need to defeat guards and rioting inmates in order to continue. If you can, perform perfect evades by pressing ⬤ / Ⓑ at the very last moment for some extra XP. After defeating everyone, you'll be able to retrieve the balloon with the book from the basketball hoop.

Reach the Belfry

Now we're ready to get back to the Belfry. Open your map and Ⓡ③ to fast travel to it.

Talk to Alfred

At the Belfry, approach Alfred left of the console to talk to him.

Open Harley's Book and Analyze Her Evidence

Now approach Harley's book sitting on the boxes near the Evidence Boards.

This finishes Case 01: Batman's Last Case in Gotham Knights.

# 02: The Rabbit Hole

## 2.1 – AKA Oswald Cobblepot

Reach the Iceberg Lounge

Leave the Belfry and proceed to the yellow marker. On the rooftop, enter the building.

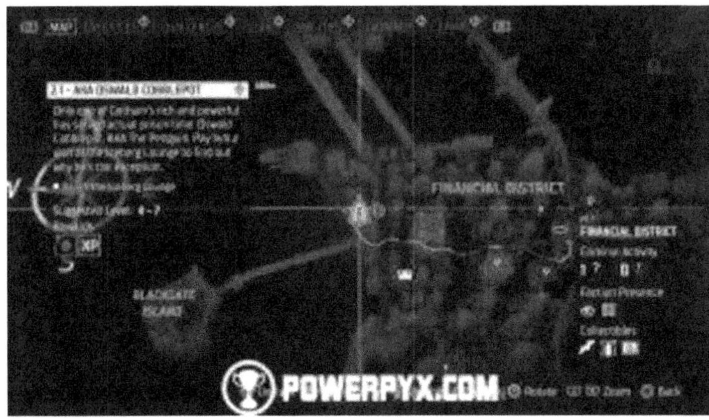


Talk to the Penguin

We're now inside the mansion of the Penguin. Go through the door.

Talk to the Penguin – Defeat the Penguin's Guards 0/8 – BONUS: Perform Silent Takedowns

Continue downstairs and get rid of all the Penguin's guards. If possible, try performing three Silent Takedowns for some extra XP.

Talk to the Penguin – Reach The Penguin's Office

Once you've cleared out the area, continue on upstairs and through the door.

Exit the Iceberg Lounge

With the Penguin not wanting to help us, we can leave the building. Follow the waypoints out of the building, not forgetting to open the chest right before the exit.

Return to the Belfry

Open up the map and R3 fast-travel to the Belfry to regroup with and inform the others.

Talk to Alfred

At the Belfry, approach Alfred to talk to him. To get the Penguin to talk, we need to get his attention some other way.

Listen to the Recorded Message

Approach the recorded message from Lucius Fox near the console and listen to it.

Check the Evidence Boards

Now approach the Evidence Boards. Hover over the Leads section and press ⬛ / ⬤ for a quick overview of the challenges awaiting us outside.

Exit the Belfry and Patrol Gotham City

When you're ready, exit the Belfry to continue on. Once outside, 🆃🅿 to open the list of available activities and ⊗ / ⬤ on The Rabbit Hole, then ⬛ / ⊗ . Here's a list of all the activities we need to complete to continue on with the story. We're going to start with The Penguin's Criminal Deal, which is the West End district.

Reach the Criminal Deal

Make your way to the location in West End.

Stop the Deal Between the Two Gangs – Defeat all Enemies 0/10 – Retrieve the Modchips 0/3 – BONUS: Perform Grab Strikes 0/3 – BONUS: Perform Perfect Attacks 0/3

At the location, defeat all enemies and then retrieve the mod chips from the briefcases. If possible, try to perform three grab strikes and perfect attacks for some extra XP. Now we're going for The Penguin's Organ Trafficking, which is in Robinson Park.

Reach the Hideout

Make your way to the location in Robinson Park.

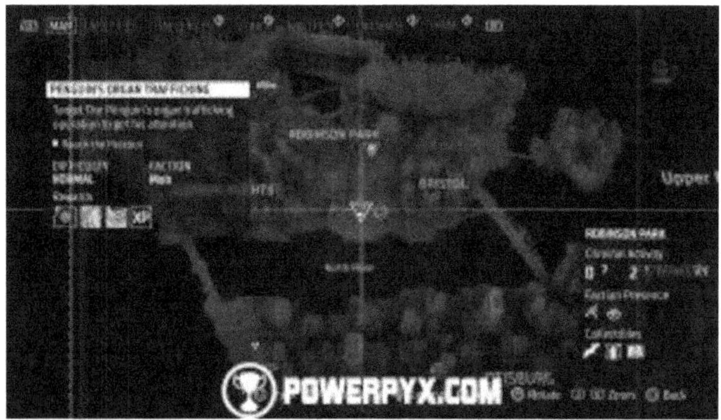

Stop the Organ Theft – Locate and Scan the Cryogenic Containers

At the location, you need to stop an organ theft, locate and scan some cryogenic containers. Hold ✛ to enter AR View and look for the white glowing item near the ambulance. Once you've identified it, clear out the area of all enemies to proceed undisturbed, paying extra attention to the cameras or disabling them if you wish.

Stop the Organ Theft – Retrieve the Cryogenic Container

Now pick it up.

Stop the Organ Theft – Reach Thompkins' Mobile Clinic

As soon as you pick it up, a countdown will start during which you need to reach the next location. So, open up the map to know where you need to go, call up your Batcycle, and head there.

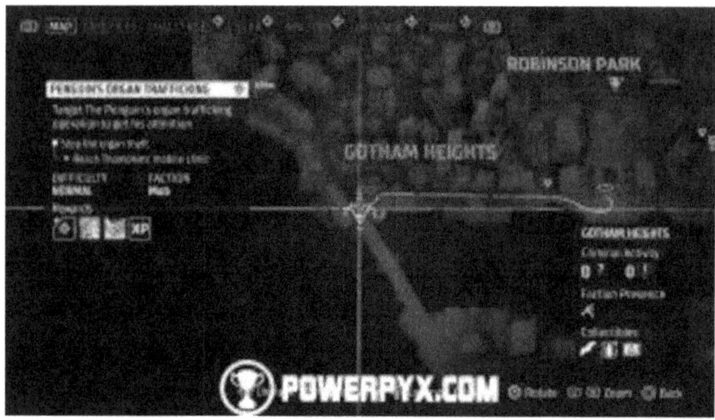

At the location, talk to Dr Thompkins. Now we're going for Lucius Fox, who is located in Otisburg.

Reach Foxteca

Make your way to the location in Otisburg.

Talk to Lucius Fox

After the cutscene with Fox, talk to him again and this will complete this challenge. Now we're going for Detective Montoya.

Visit Detective Montoya

Make your way to the location in Old Gotham.

Reach the Iceberg Lounge

Now we're ready to head back to The Penguin's mansion.

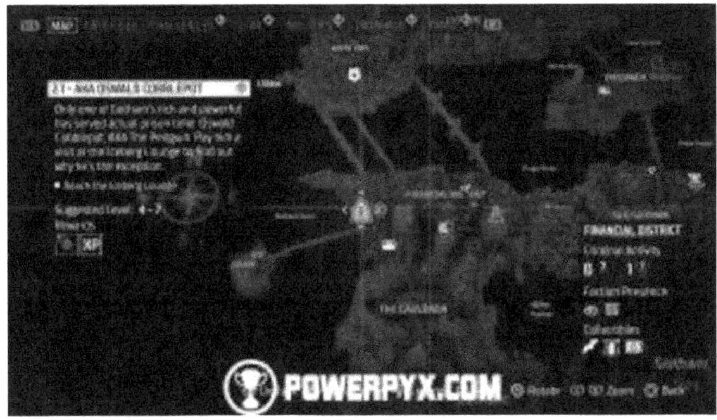

Enter the Iceberg Lounge

Once on the rooftop, enter the lounge.

Talk to the Penguin

As before proceed through the doors.

Talk to the Penguin – Defeat the Penguin's Guards 0/7 – BONUS:
Avoid Taking Damage

In the new room, defeat all of Penguin's Guards. If possible, stay out of harm's way for some extra XP.

Talk to the Penguin – Reach The Penguin's Office

Once you've cleared out the area, continue on upstairs and through the door.

Locate and Destroy the Bugs 0/3

Someone's bugged the Penguin's office. We need to find the bugs and destroy them. One bug is below the Penguin's bust, another one is behind the Penguin himself, and the third one is by the lamp on his desk.

Pick Up the Bottle of Whisky

Now approach the liquor cabinet and grab the bottle of whisky.

Exit the Iceberg Lounge

We now have a new heading. Head downstairs and exit the Iceberg Lounge

## 2.2 – The Powers Club

Check on Alfred and his Evening Meeting

Head to Alfred's location in Tricorner Island. Reach the rooftop for a new cutscene.

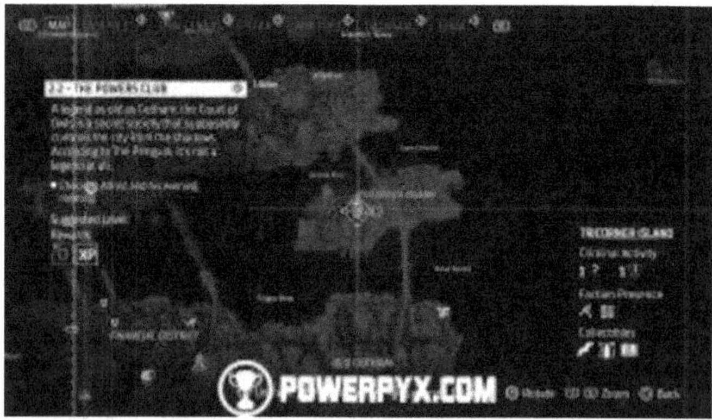


Reach Powers Club

Now head to the Powers Club in Old Gotham. Enter the greenhouse through the rooftop.

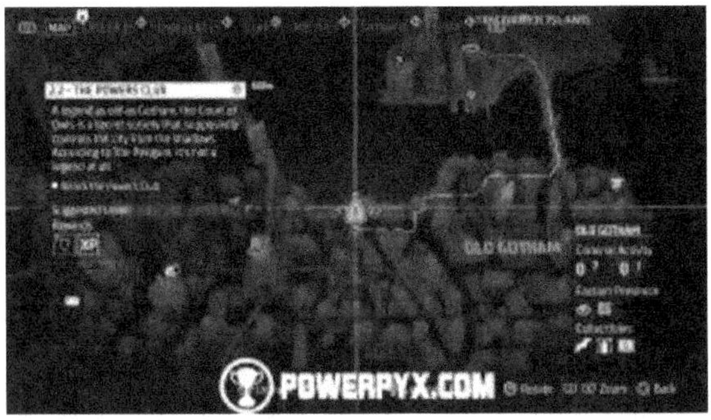

Enter Powers Club

Now proceed through the doors.

Enter the Club

Continue along the hallway to access the Club.

Defeat the Security Guards

Now drop down and defeat the two guards.

Use AR to Scan the Area

Now press ✥ to activate AR and scan the yellow trails on the floor.

Scan the Hidden Mechanism

Next, scan the red hidden mechanism under the floor.

Find a Way to Activate the Mechanism – Follow the Wires

After scanning the mechanism, you should now have wires highlighted in your AR view. Follow them by going upstairs and opening the door.

Find a Way to Activate the Mechanism – Clear the Guards from the Club – Use AR to Scan for Clues – BONUS: Remain Undetected

Now continue along the hallway and into the enemy-ridden area. Here, clear out the enemies. If possible, stay undetected for extra XP. Then, as if you were entering the place for the first time, scan the bust found inside the left room.

Find a Way to Activate the Mechanism – Press the Button

Now press the button on the bust. That's one switch taken care of. Now we need to find the other.

Find a Way to Activate the Mechanism – Use AR to Find the Other Switch

Next, go to the other room and analyze the book in the corner.

Find a Way to Activate the Mechanism – Flip the Switch

Now press the button inside the book. That's the second switch taken care of.

Find a Way to Activate the Mechanism – Follow the Activated Wires

Return where you've fought the two guards and activate the lamp from the wall.

Follow the Blood Trail

Head down the secret stairs and open the door.

Continue Following the Blood Trail

Continue to the nest room and approach the paintings on the wall.

Use AR to Scan for Clues

Now scan the spotlight at the start of the room.

Activate the Spotlight

Activate the spotlight to reveal a yellow wire and several eagle statue pieces that if looked at from a certain angle align with the shadow on the painted walls.

Solve the Puzzle

The goal of the puzzle is to rotate the pieces to drop a shadow on the wall that looks like a bird facing to the right. The "shadow bird's" head and talons must face to the right side, not the left.

If you haven't interacted with any tables yet, the solution is as follows (you can pause the game and return to Main Menu, then press Continue to reset the Puzzle):

Figure 1 (Talons) = 1 Click

Figure 2 (Beak) = 2 Clicks

Figure 3 (Wing 1) = 3 Clicks

Figure 4 (Wing 2) = 3 Clicks (it may take a fourth click to trigger the cutscene)

The painted wall should now open up – see the finished shadow bird below:

Defeat the Court of Owls Guards 0/4

Defeat the guards.

Find where the Blood Trail Leads

Head downstairs and open the door, then grapple up and onto the balcony above.

Defeat the Court of Owls Guards 0/5 – BONUS: Perform Silent Takedowns 0/3

Clear out the area of enemies. If possible, perform Silent Takedowns for extra XP.

Defeat the Remaining Guards 0/8

Now defeat the next set of guards coming your way.

Continue Following the Blood Trail

Now we can resume following the blood trail. Open the central doors.

Follow the Blood Trail

Continue following the blood trail through the lairs until you see a man tied to some owl's wings.

Untie the Prisoner

Approach the man and untie him to save him and trigger a new cutscene.

Escape the Trap

We now need to escape a trap in true Riddler's fashion, avoiding blades and fire. Just run through the traps and time your movements well to not get killed.

Find a Way to the Surface with the Key

Continue through the cave until you eventually can squeeze through a wall.

Return to the Belfry

Time to get back to base. So, open up your map and 🄬 to return to the Belfry.

This finishes Case 02: The Rabbit Hole in Gotham Knights.


## 03: In the Shadows

### 3.1 – The Key

Reach the Gotham Gazette

Exit the Belfry and make your way to the next destination, the Gotham Gazette HQs.

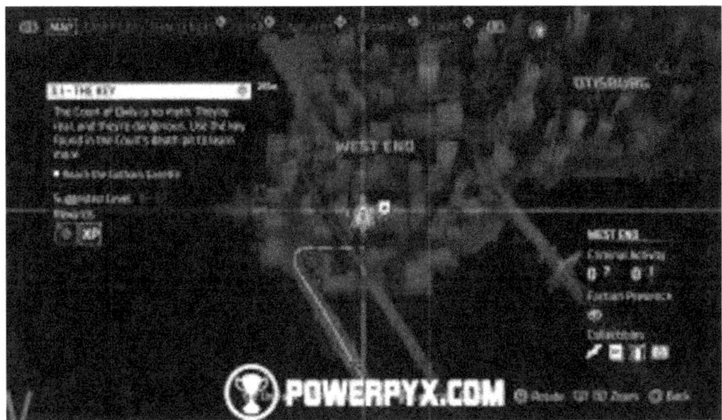

## Find the Door

Get to the rooftop, exactly where the yellow marker directs you to find the door to the hideout. Then, approach the wall to open it.

## Search Inside the Owl's Nest

Enter the Owl's Nest, get downstairs and approach the wooden cabinet.

## Scan the Document

Next, enter AR View and scan the document.

Investigate the Table

Nest, approach the wooden table nearby and hold ⊗ / Ⓐ to investigate.

What is the Order about?

For this objective, we need to determine the Court's instructions and the target location. For the solution, select the Ofásiede Card (sword car) and the brass peg (fourth pin on the map from top to bottom).

Exit the Court Hideout

Retrace your steps and leave the hideout.

## 3.2 – Chelsea Tunnel

Reach the Chelsea Tunnel Construction Site

Proceed to your next objective.

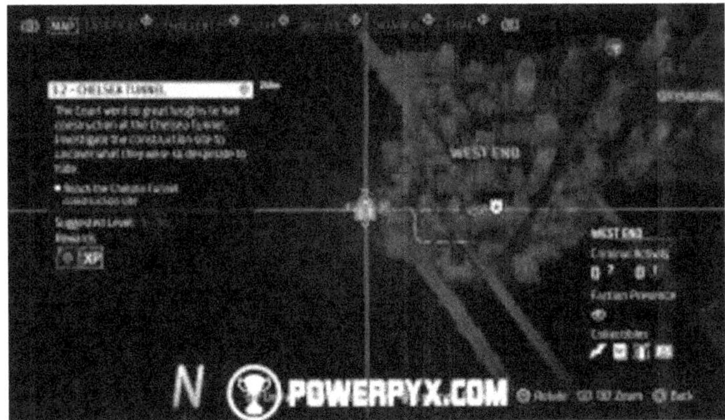

Reach the Entrance

Now enter the tunnel between the two statues and when you're in front of the locked gate, look up and to the left to find an opening through a lit air vent fan.

Now drop down and ⊗ / Ⓐ in front of the hole in the ground to continue on.

Search the Tunnels

Drop down into the tunnels and continue on until a new cutscene revealing a new enemy type unlocks.

Defeat the Talon

The Talon enemy type is really good at avoiding your standard melee attacks, so they won't work. However, Heavy Ranged Attacks will. When successfully performed, this should stun the Talon, allowing you to then to hit him with standard melee attacks.

Search the Cave

Continue through the tunnel and open the door.

Reach the Other Side of the Chasm

We're now in a huge chasm and we need to reach the other side. Head left past the gap, then immediately go right and grapple to the balcony on the right side to find a chest. Now grapple to the wooden scaffolding to your right to find another chest. Now continue on towards the white waypoint.

Explore the Old Mine

Continue on to reach a mine until a new cutscene unlocks.

Defeat the Talons in the Pit

More Talons! Use the strategy explained above to defeat them all.

Find a Way Out of the Pit

Continue on until you get to a split. Go right first to find a chest, then open the door.

Reach the Main Level

Grapple to the structure above.

Search the Main Level

Continue on and drop down the balcony.

Search the Main Level – Find Excavation Research 0/3

We now need to find three samples of excavation research. The first one is in the bulleted wall, the second one is up the elevator shaft on a table with two spectrographic analyzers, and the third one is at the end of the room, near two x-ray analyzers and a chest.

## Search the Main Level

Now go back to where you've found the second sample and interact with the keypad next to the green door. Continue up the main path, up the elevator shaft, jumping from platform to platform.

Investigate the Extraction Process

When this new objective unlocks, drop down below. First off, open the chest in the back near the metal containers. Then, approach the table to start investigating.

Extract a Viable Sample

For this objective, we need to determine which sample should be placed in the extract and the chemical that will extract material from it. For the solution, select Composite Ore and the orange container.

Exit the Extraction Site

Go back to the upper level and exit through the door.

Reach the Elevator

Continue on to reach the elevator.

Activate the Elevator

Approach the console to activate the elevator. However, more Talons get in our way.

Defeat the Talons

Defeat the Talons, paying extra attention to their flurry attacks this time as they infect you with poison on hit.

Escape the Collapsing Mines

The elevator got destroyed and then caused the mine to start collapsing. Follow the white waypoints, grappling from platform to platform and ignore any enemy that may come in your way.

Search for a Way to the Surface

Continue on the collapsed tunnel until you see a locked door. Now, turn your camera 180° and grapple above.

Keep going, grappling higher and higher until you can climb up the stairs. Behind the stairs is a chest. Then, exit back to Gotham through the blue door.

Return to the Belfry

As soon as you exit the tunnels, you'll be tasked with going back to the Belfry. However, the connection with the Belfry breaks up and you're tapped by Talia al Ghul.

Talk to Talia al Ghul

Proceed to Talia's location in the Financial District.

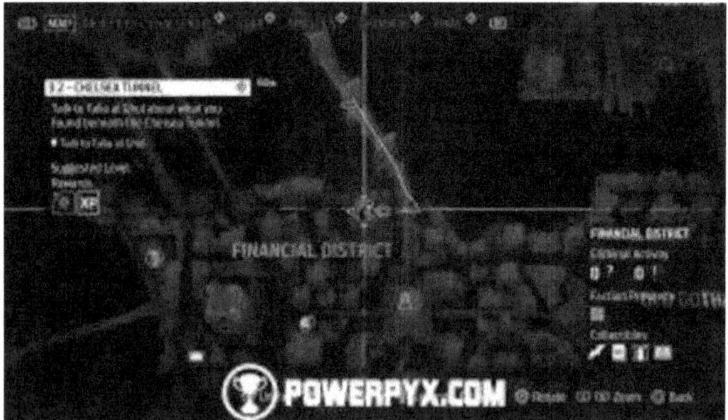

Return to the Belfry

After talking to Talia, we can actually head back to the Belfry. So, open up your map and 🅡🅢 to fast travel to it.

This finishes Case 03: In the Shadows in Gotham Knights.

# 04: The Masquerade

## 4.1 – Mark Hendricks

Check the Evidence Boards

Approach the Evidence Boards to examine the new evidence.

Reach Mark Hendricks' Last Known Location

Exit the Belfry and make your way to the main yellow marker in south Tricorner Island. When you get to the location, wait around the area for 10-15 seconds until the new objective unlocks.

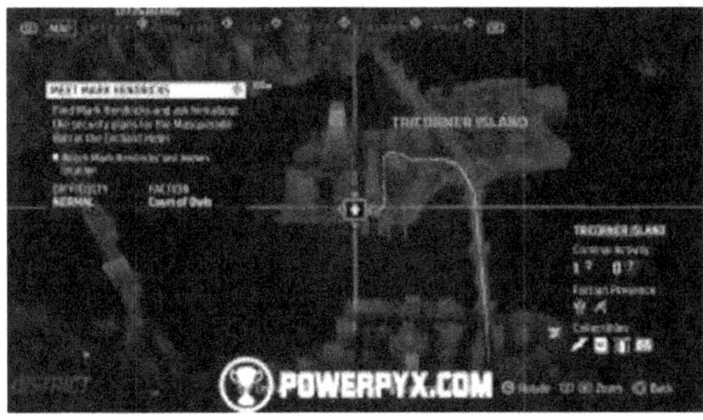

Locate Mark Hendricks

Now head to the new yellow marker in Tricorner Island.

Protect Mark Hendricks – Defend the Doors 100% – Defeat Wave 0/3 – BONUS: Defeat Enemies with Perfect Attacks 0/3

At the new location, drop down and start clearing out the enemies. You'll have to defeat three enemy waves before the doors' percentage drops down to 0. If possible, try to land three Perfect Attacks for some extra XP.

Talk to Mark Hendricks

Once the area is clear, Mark Hendricks will immediately come out the door, so you can talk to him.

Return to the Belfry

As usual, open up your map and **R3** to get back to the Belfry to rendezvous with the others and get ready for your next step.

## 4.2 – The Orchard Hotel

Enter the Orchard Hotel

Leave the Belfry and make your way to the next yellow marker, the Orchard Hotel in Otisburg. The entry point is on a rooftop of the hotel itself.

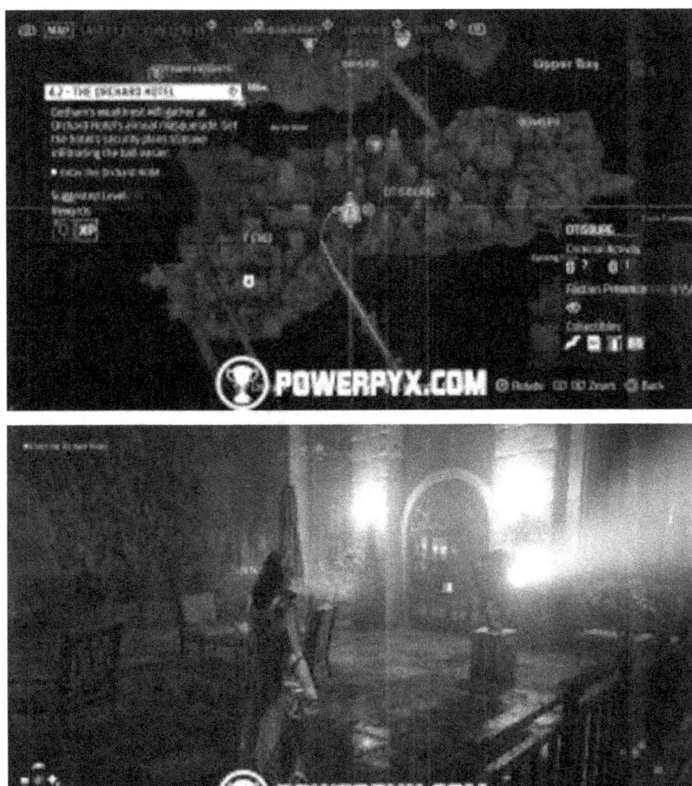

Reach the Ballroom – Avoid Detection

We need to head for the ballroom, but can't get spotted. Crouch through the hallway until you get to the buffet area. Then, continue crouching along the wall until you get to a corner.

Here, stop and wait for the Court guard to walk away first.

When he moves, continue behind the couches and wait for him to leave behind the second pillar.

Keep following the white waypoint until you can open a door.

Reach the Security System – Avoid Detection

Now we need to reach the security system and disable it while still avoiding detection. Get past the first camera and around the corner, then stop. You should see a group of Court guards that can be easily snuck past as they're not looking our way.

Before heading inside the security system room, make a slight detour to the left for a chest.

Now, enter the security system room.

Disable the Security System – Avoid Detection

Silently take out the guard and then disable the security system.

Defeat the Security Guards

Now defeat the guards outside.

Reach the Ballroom

Go where the guards were gathered and the camera will point you to an opening above you can grapple to, then open the gate.

## Eavesdrop to Identify Court Members

Here you need to eavesdrop on specific people. The people are: the lady in the wheelchair in the middle of the room, the two guests at the balcony above the lady in the wheelchair, and lastly the man with an owl mask in front of Bruce's portrait photo showing up last. The man in the owl mask might not show up until you eavesdropped on all available conversations.

Locate the Voice – BONUS: Remain Undetected

Make your way back by means of the open-air vent. Go through the door until you see more Court guards. Defeat them all. If possible, don't get caught for some extra XP.

Locate the Voice

Once you've gotten rid of everyone, there are two chests in the area which can be found using AR scan. Continue down the main hallway until you see an air vent as marked by a white waypoint.

Reach the Voice's Location

Climb up and through the air vent.

Find the 13th Floor

Interacting with the piano as the Voice did won't work, so head through the doors. Continue down the hallway until you can open an Exit door leading to the stairwell. Keep going downstairs until you a secret door. Before continuing, keep going downstairs for a chest.

Open the Secret Door

Now interact with the secret door to open it.

Search the 13th Floor for the Voice of the Court

Go down the hallway until you see a fireplace. Then, scan the place to find an owl statue sitting opposite the fireplace that you can interact with. Approach the statue and press the switch to reveal a secret hallway.

Defeat the Court Members

Continue down the hallway and three new Court members will show up. Get rid of them.

Locate Evidence

Now scan the touchscreen one of the Court members was interacting with.

Activate the Projector

Lastly, activate the projector by interacting with the switch next to the touchscreen.

Insert the USB Drive

Interact with the touchscreen again to insert a USB stick and get the evidence.

Remove the USB Drive

Now remove the stick. We're done here.

Defeat the Court Members

Now defeat the two Court members.

Search for a Way Forward

Exit through the moonlit room the two Court members came out from to access a puzzle.

Solve the Puzzle

Here you have a large floor map with 4 Gotham buildings around it. To solve the puzzle, you need to step on the plates on the floor in the correct order, from oldest to newest. The correct order is given to you by the years found under the paintings in the same room. From the way we came in, we're going to number the buildings 1 to 4, left to right. Now step on: 4-2-1-3. Solving the puzzle reveals a new secret compartment on the left side of the room.

Search the 13th Floor for the Voice of the Court

Walk along the hallway until a new door opens up in front of you.

Talk to the Voice

Enter the room and approach the painting on the left to trigger a new cutscene.

Defeat the Guards 0/8

New guards incoming! Time to get rid of them.

Return to the Ballroom

Now continue through the hallways and up the elevator shaft to head back to the ballroom.

Defeat Enemies 0/3

As you make it back to the ballroom, you'll be stalled by some enemies, including a new enemy type called the Assassin, who, for instance, can be defeated with elemental attacks.

Return to the Ballroom

Now resume your run back to the ballroom and drop down below.

Defeat the Assassins

In the ballroom you'll have to fight League Assassins using a cannon to shoot at you. Avoid standing in the red circle when they're about to shoot at you and land your attacks when you have a window of opportunity. If they disappear, stand still and when they reappear, quickly dodge their attacks.

Reach the Belfry

As usual, open up your map and **R3** to return to the Belfry.

This finishes Case 04: The Masquerade in Gotham Knights.


## 05: The Court of Owls

### 5.1 – Little Birds

Talk to Talia al Ghul

Exit the Belfry and proceed to Talia's location in the Financial District.

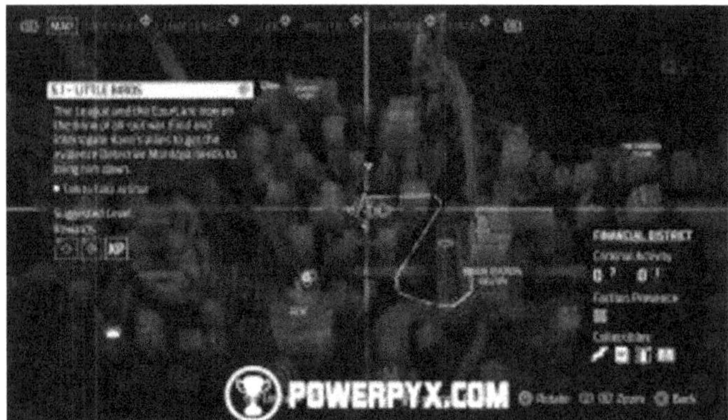

Talk to Detective Montoya

Leave the garage and proceed to Detective Montoya's location in Old Gotham.

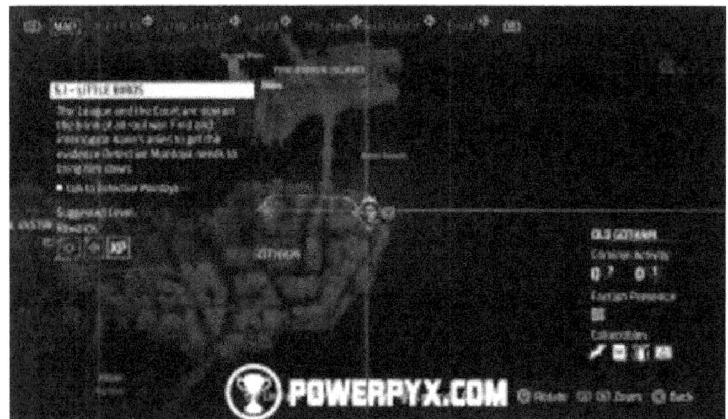

Stop Corrupt Detectives 0/5

For the next step, we need to find 5 corrupt detectives and eliminate them. Here are two possible locations.

Reach the Hideout

Now make your way to the hideout's location in Bristol.

Reach Gotham City Cemetery

When you get to the location, enter the cemetery.

Reach the Hideout

Continue making your way down to the hideout.

Disrupt the Court of Owls' Plan – Defeat all Enemies 0/5 – BONUS: Perform Throws 0/2

At the location, defeat all enemies. If possible, perform 2 throws.

Disrupt the Court of Owls' Plan – Find and Scan Evidence of a Cover-up

Now scan the Owl's mask near the barrel on fire.

Return to the Belfry

As usual, open up your map and R3 to fast-travel to the Belfry.

Listen to The Penguin's Recorded Message

Approach the recorded message by the console to listen to it.

Reach the Iceberg Lounge

Leave the Belfry and head out to the Iceberg Lounge in the Financial District, as usual, entering from the rooftop.

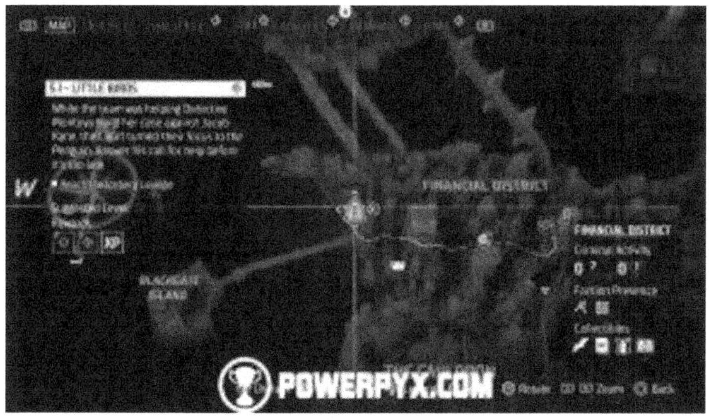

## 5.2 – Inside Gotham's Walls

Talk to the Penguin

Head upstairs to enter the Penguin's office. The Penguin was pinned to the wall.

Escape the Labyrinth

You're now in a labyrinth with several different traps and every time you die, you start over. Run through the labyrinth and when you see holes in both walls, crouch to avoid getting killed by the trap.

Continue on and when you get to a room with overhead torches, you have to walk under those that are off to make it to the other side.

Continue on until you reach a dead-end, then turn around and keep going until you reach a room with moving spikes for a checkpoint. Here time your movements carefully to make it to either exit.

For the next few sections, choose either path and you'll eventually reach the exit.

Find a Way Out of the Room

You're now in a room with a big owl-looking sphere in the middle. In order to solve this puzzle, you have to quickly step on the four pads found in the room. From where you came in, there's on your left, on the right across the room, one behind and above the sphere, and one above the entrance to this room. The timing here is quite lenient, so choose what you think is the fastest way to turn them all on and then proceed through the door. If you fail, you'll have to defeat some Talons and then try again.

Escape the Labyrinth

We're now back in the labyrinth. Continue down either path until you'll have to fight some more Talons.

Escape the Labyrinth – Defeat the Talons

Beat all three Talons to continue on.

Escape the Labyrinth

Continue on past the statues.

Escape the Labyrinth – Defeat the Talons – BONUS: Defeat Enemies Using Momentum Abilities

Time to defeat some more Talons. If possible, use Momentum Abilities for some extra XP.

Escape the Labyrinth

Now continue past the door that opens and continue on until you unlock a new cutscene.

Defeat the Gladiator Talon

Unlike normal Talons, the Gladiator Talon doesn't avoid your standard melee attacks. However, he's got a shield, so you must use heavy melee attacks to find an opening for normal melee attacks. Continue doing so until a new cinematic unlocks.

Search the Staging Area

Continue past the collapsed statue, then drop down. Grapple to the platform in front of you and make a slight detour to the left for a chest, then continue on the opposite way.

Clear the Court Operators from the Control Room – BONUS: Remain Undetected

First, grab the second chest in the area and then approach the control room using the railings above.

From here, drop down and kill the enemies. Alternatively, you can try landing some Silent Takedowns for some extra XP.

Defeat the Enemies

Kill the three Court operators and the incoming Talons.

Use the Room Assembly Controls

Now approach the wheel on the console to use it.

Reach the Platform

Now, as the stage moves, reach the first platform in front of you.

Scan the Fuel Tank

From here, enter AR View and scan the big red fuel tank in front of you. Left of the tank is a chest.

## Break Open the Thermal Valves 0/2

Now get closer to the fuel tank and use a Batarang to aim at and destroy the thermal valves on its side. This will blow the fuel tank up, allowing us to continue on.

Escape the Room

Now grapple onto the fence and drop down the hole.

Find a Way Out of the Underground

Continue along the underground tunnels, making sure you avoid the ice jets along the way.

Reach the Court Laboratory

At the end of the tunnel, grapple onto the ledge above and open the chest at the end of the section.

Now retrace your steps and enter the opening halfway through.

Interrogate the Head Surgeon

From the ledge, perform an Aerial Attack on the head surgeon, weaken him and **R2** / **RT** to grab and interrogate him when the prompt appears.

Open the Gate

Exit the laboratory by opening the gate at the end.

Activate the Elevator

Now activate the elevator by interacting with the switch.

Clear the Court Operators from the Coffin Room

Now clear out the area of all enemies.

Defeat the Enemies

Now repeat the same on the Talons.

Override Hibernation Controls 0/2

Approach the wheels on the consoles to override the hibernation controls. As you do, more Talons will show up.

Defeat the Talons 0/3

Now get rid of the final group of Talons.

Override the Failsafe

Lastly, override the failsafe from the central console.

Defeat the Gladiator Talons 0/2 – BONUS: Perform Perfect Evades 0/5

Now defeat the 2 Gladiator Talons. If possible, perform 5 Perfect Evades for some extra XP. Remember that a Perfect Evade takes place when you press ⬤ / Ⓑ at the very last moment.

Escape the Area

Time to leave this place.

Return to the Belfry

As usual, open up your map and 🄡 to fast-travel to the Belfry.

This finishes Case 05: The Court of Owls in Gotham Knights.


## | 06: Jacob Kane

*6.1 – Court Judgement*

Talk to Detective Montoya

Leave the Belfry and head out to Detective Montoya's location in Old Gotham.

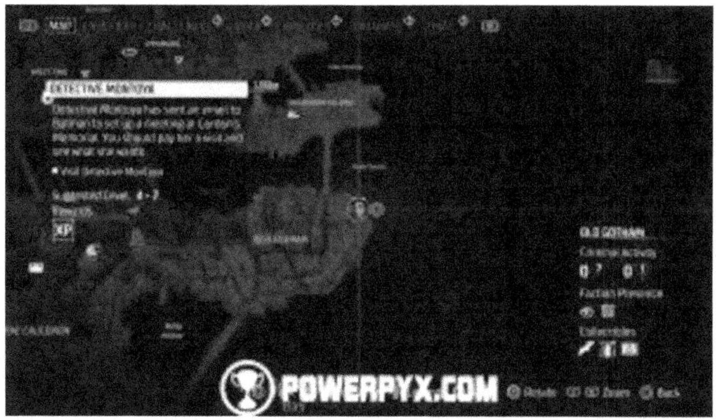

Interrogate The Mob to Find Out More Information

We now need to locate a mobster to find out more information in the West End district. Here's a possible location.

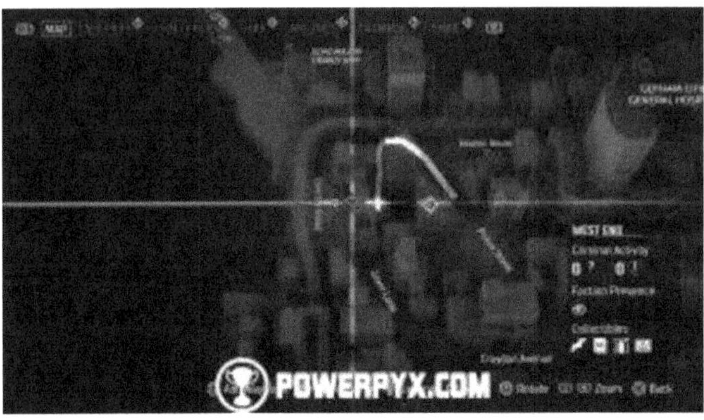

As usual, weaken the enemy until you can **R2** / **RT** grab and ⃝ / ⃝ interrogate him.

Interrogate Regulators to Find Out More Information

Now we need to locate a regulator to find out more information in The Cauldron district. Here's a possible location.

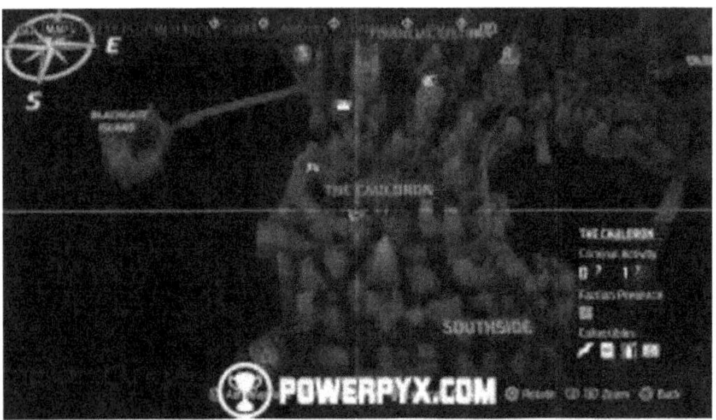

As usual, weaken the enemy until you can **R2** / **RT** grab and ⃝ / ⃝ interrogate him.

Interrogate Freaks to Find Out More Information

Now we need to locate a Freak to find out more information in Robinson Park. Here's a possible location.

As usual, weaken the enemy until you can R2 / RT grab and ▲ / Ⓨ interrogate him. If for some reason you can't find any ongoing Freak crimes, return to the Belfry to end the patrol and try again.

Reach Judge Moreno

We finally have Judge Moreno's location in Bristol.

Protect Judge Moreno – Defend the Door 100% – Defeat Wave 0/3 – BONUS: Perform Silent Takedowns on Marked Targets 0/1 – BONUS: Perfect Throws 0/3

At the location, you have to defeat three enemy waves while making sure that the door's integrity doesn't reach 0%. If possible, perform one Silent Takedown on the marked targets and 3 Perfect Throws.

Talk to Judge Moreno

Once the area is clear, Judge Moreno will come out of the door for you to talk to her.

Reach the Belfry

Time to head back to the Belfry. As usual, open up your map and R3 to fast travel to it.

*6.2 – The Voice of the Court*

Inspect the Kane Drone

Inside the Belfry, inspect the Kane Drone right of the console where the white waypoint is. This will teleport you outside, automatically starting this sub-case.

Reach the Kane Industries Rooftop

The Kane Industries Building isn't exactly grapple-friendly, so we need to find an alternative way to climb it. From the water tower, grapple to the budding in front of you.

From here, grapple to one of the ledges of the building.

Next, shimmy rightwards and grapple higher to the other side.

Now go around the building until you see a platform you can grapple onto above you.

Now press ⊗ / Ⓐ as you grapple to jump and use the jump momentum to reach the top of the building.

Defeat the Guards

On the rooftop, defeat all enemies

Enter the Building

Now get to the very top and interact with the console to access the building.

Use AR to Locate an Experimental Laser Drill

We need a laser to continue through. Activate AR View and go where all the white crates are. Here, in a corner, is also the laser drill.

Take the Laser Drill

Now collect the laser drill.

Defeat Kane's Security Officers 0/4 — BONUS: Defeat Enemies Undetected

Defeat all enemies. If possible, do it while being undetected for some extra XP.

Pick Up the Access Card

Now pick up the access card from one of the guards' bodies.

Search for a Security Station

Enter the hallway the guards came out from and continue on until you spot an opening above. Climb it up and continue on and down onto the security station.

Rewrite the Access Card

Now access the console and rewrite the access card.

Reach the Executive Floors

Leave the security station through the door and before going through the door with the white waypoint, open the other one to find a chest.

Now continue on with the main door and all the way downstairs.

Defeat Lobby Guards 0/10

Get rid of the guards.

Deactivate the Alarm – Enter Jacob Kane's Office – BONUS: Remain Undetected

Go through the door and open the chest immediately to your left, then deactivate the alarm from the console. This will start a countdown. Now quickly follow the white waypoint and get to the other side before the time is up.

If you didn't get caught by the security cameras, you will have completed the bonus objective.

Search for Kane's Private Elevator

As soon as you enter Kane's office, open the chest to your left.

Then, activate AR View and interact with the wall opposite the writing desk in the back of the room.

This is a sudoku puzzle. The solution is in the screenshot below.

Now go through the doors and open up the chest immediately to your left and then the one by the fireplace.

Now enter AR View and scan the globe near the painting and interact with it.

Use the Laser Drill on the Elevator Door

Approach the elevator door now and use the laser drill on it. However, the noise of the laser drill will attract some goons.

Survive Until the Laser Drill Finishes 0% – BONUS: Avoid Taking Damage

Now you need to defeat some enemies as the laser drill does its thing. If possible, avoid sustaining damage for some extra XP. At some point, the laser drill will get stuck and you'll need to reactivate it.

Survive Until the Laser Drill Finishes 42% – Reactivate the Laser Drill

When the laser drill gets stuck, approach it to reactivate it.

Survive Until the Laser Drill Finishes 42%

Now simply defeat all remaining enemies.

Defeat Kane's Security Officers

As soon as the elevator door opens, new enemies will come your way for you to defeat.

Enter Kane's Private Elevator

Enter the elevator to continue up.

Search the Bunker for Jacob Kane – BONUS: Perform Silent Takedowns 0/4

As soon as you're done riding the elevator, crouch and get rid of all guards. If possible, perform 4 Silent Takedowns for some extra XP on the guys in the corridors.

Search the Bunker for Jacob Kane – Defeat the War Room Guards 0/3

Now get rid of the guards around the table.

Search the Bunker for Jacob Kane

Continue downstairs and open the red door.

Defeat Kane's Security Officers 0/4 – Disable the Security Turrets – BONUS: Remain Undetected

Time to defeat more security officers. This time, though, you also have to disable some security turrets. If possible, do all of this without getting caught for some extra XP. To disable the turrets, you need to access the panels near the garage door and the Inclinator.

Activate the Inclinator

Now activate the Inclinator at the panel to continue down.

Survive the Descent – BONUS: Avoid Taking Damage

As the Inclinator descends, new enemies will come attack you. Defeat them all and try to avoid getting damaged for some extra XP.

Defeat the Talons 0/3 – BONUS: Defeat Enemies Using Momentum Abilities 0/3

When the Inclinator stops, defeat the three Talons waiting for you. If possible, defeat them using Momentum Abilities for some extra XP.

Search for Jacob Kane

Continue through the next two doors until you reach a new enemy area.

Search for Jacob Kane – Defeat Kane's Security Officers 0/5 – BONUS: Defeat Enemies Using the Environment 0/2

Eliminate all enemies here. If possible, kill 2 of them using environmental opportunities (highlighted in red and can be activated with a Batarang)

Search for Jacob Kane – Unlock Door to Reactor Room

Once the area is clear, access the console to activate the door to the reactor room.

Search for Jacob Kane – Enter the Reactor Room

Now walk through the door to enter the reactor room.

Search for Jacob Kane – BONUS: Find and Open Chests 0/4

In this area there are 4 chests to open needed to complete the bonus objective. Once you reach the room, get past the poison gas to find the first.

Now continue on and turn around where the fire hazards are to find the second.

Now get through the fire hazards to get back to the beginning of the area and grapple onto the balcony above for the third chest.

Now continue on the main path and you'll find the fourth and final one in your path.

Search for Jacob Kane

Continue upstairs and open the red door for a new cutscene.

Find a Way to Breach Blast Door – Clear Out all Court Members in the Bunker 0/10 – BONUS: Perform Silent Takedowns 0/5

Clear out the area. If possible, perform 5 Silent Takedowns for some extra XP. There are also two chests in this area for you to open and they can easily be found using AR View.

Find a Way to Breach Blast Door – Scan the Blast Door

Now scan the blast door.

## Find a Way to Breach Blast Door

Grapple above the blast door to interact with a console that moves the laser. Press Move Laser Up once and Move Laser Left once. Now proceed to the other console, the one moving the platform, and press the switch three times. Now get back to the other console and Move Laser Down.

Find a Way to Breach Blast Door – Activate the Laser

Now return to the other console and activate the laser to blast the door open.

Chase Jacob Kane

Get through the blast door, open the chest, and then force the red door open. Continue running downstairs, past all traps, and open the door for a new cutscene.

Defeat Hunter Talons 0/2 – BONUS: Perform Perfect Attacks 0/4

Defeat the two Hunter Talons. If possible, perform 4 Perfect Attacks for some extra XP.

Stop Jacob Kane

Continue through the doors until you unlock a new cutscene. Kane is getting arrested.

Chase Talia al Ghul

Chase Talia's glow through the rooftops. If you lose track of her, look for white waypoints along the way.

## Reach Wayne Tower

After a while, you'll have to reach Wayne Tower. Get to the top for a new cutscene.

## Defeat the League of Shadows Assassins

Defeat the Assassins. Remember that to defeat them you have to wait for them to appear out of nowhere, then avoid their ninja attacks and only then can you damage them.

After defeating them, you'll automatically get back to the Belfry.

This finishes Case 06: Jacob Kane in Gotham Knights.

## 07: The League of Shadows

### 7.1 – Friends In Need

Exit the Belfry and Patrol Gotham City

Leave the Belfry. As soon as you get out, Alfred will inform you that Lucius Fox and Detective Montoya have been kidnapped by the League of Shadows. We first need to rescue both in order to continue. Let's start with Lucius Fox.

Reach Lucius Fox

Make your way to the mission marker in Otisburg and reach the rooftop to activate the next objective.

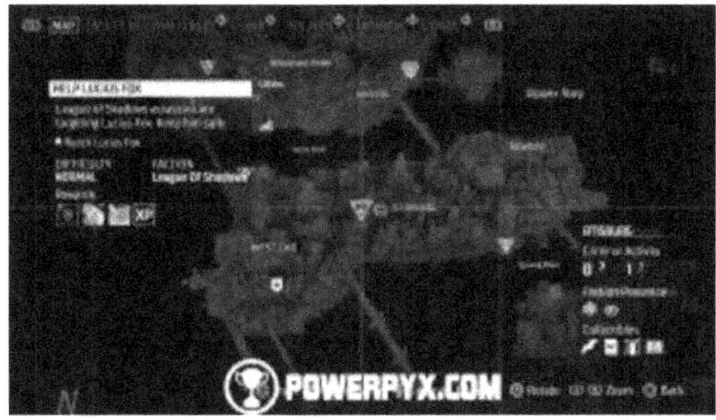

Defend Lucius Fox – Stop the League of Shadows from Shutting Down the Force Field – Stop the Hack 0% – Defeat all Enemies – BONUS: Defeat Enemies with Perfect Attacks 0/2

When you reach the rooftop, drop down and start killing enemies to prevent them from shutting down the force field. You need to stop the hack before it reaches 100%. If possible, kill 2 enemies with

Perfect Attacks of some extra XP.

Talk to Lucius Fox

Now approach Lucius Fox to talk to him. Now we can take care of Detective Montoya.

Reach Detective Montoya

Head to Detective Montoya's location in the West End district.

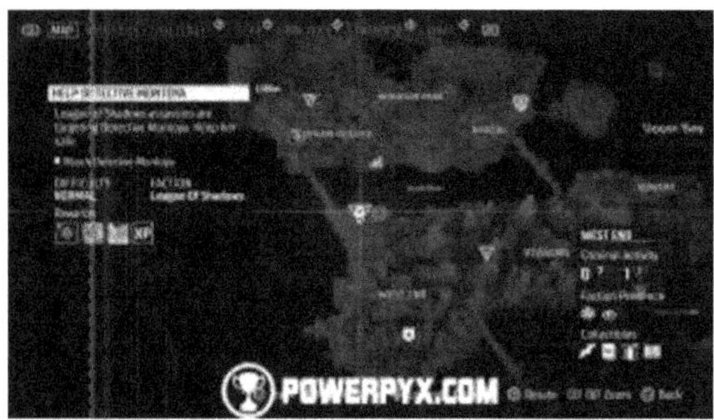

Save the Officers – Defeat all Enemies 0/4 – BONUS: Defeat Enemies with Perfect Evades 0/3

Drop down into the fight and defeat all enemies. If possible, try to score 3 Perfect Evades for some extra XP.

Reach the Belfry

As usual, open up your map and R3 to fast-travel to the Belfry.

## 7.2 – Talia al Ghul

Reach Arkham Asylum

Exit the Belfry and make your way to Arkham Asylum east of Bristol.

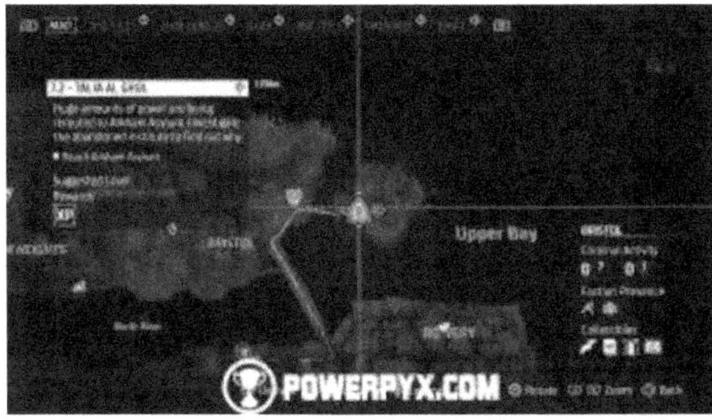

## Enter Arkham Asylum

Approach the main gate to enter it.

## Enter the Asylum

Continue through the garden and enter the Asylum.

## Reach the Solarium

Go upstairs to reach a lab.

Scan Containment Pod

Now scan the pod with AR View.

Investigate the Laboratory

Now approach the table by the blackboard to start investigating.

What Was Talia Doing in This Lab?

For this objective, we need to determine how Talia advanced Langstrom's research and was Talia was trying to achieve. The solution is Genetic Marker Group M and Serum v24.07 (pink vial).

Search for the Maximum Security Wing

Leave the lab using the opening on the left.

Continue along the main hallways and downstairs. Then drop down into the elevator shaft and open the door.

Activate Backup Generators

Enter the side room on your left to find a chest.

Now continue past the prison cells and enter the control room in the middle to pull a lever.

Return to the Main Hall

Retrace your steps now to the main hall and open the wooden door leading back outside.

Return to the Asylum

Now we need to find a way to get back inside, but the League knows we're here.

Defeat the League Assassins – BONUS: Perform Perfect Evades 0/3

Defeat the League Assassins. If possible, try to perform three Perfect Evades by avoiding attacks at the very last moment for some extra XP.

Return to the Asylum

Head to the back of the building for a chest, then get back inside.

Reach the Upper Floor

We're now back where the lab was. Go upstairs and enter the elevator.

Now grapple onto the opening above and continue to the door.

## Find a Way to Reach the Tower

Open the chest in one of the side rooms and then continue through the hallway until you get to a room with a hole in the ceiling.

From here, grapple all the way up to the tower.

## Reset the Sonic Device

Now approach the device to reset it.

Align the Frequencies

Now you need to align some frequencies. If you notice, around the tower are some amplifiers (not the ones looking outside) showing specific symbols. These amplifiers connect to the ones facing outside through colored wires. You have to follow each colored wire and set the corresponding symbol on the amplifiers looking outside. However, you must do this in a specific order. Facing the amplifiers looking outside, we're going to number them 1-4, top to bottom, left to right. Interact with 2, 4, 1, 3.

Return to the Laboratory

Drop down the tower and follow the white waypoints back to the laboratory.

Inspect Broken Pod

Now approach the broken green pod to investigate it.

Defeat the Man-Bat

To defeat the Man-Bat you have to avoid his swoop and wing attacks. Then, you can land your standard or heavy melee attacks.

This finishes Case 07: The League of Shadows in Gotham Knights.

### 8.2 – The Lazarus Pit

Reach Gotham Cemetery

Now head for Gotham Cemetery in the northern part of Bristol.

At the cemetery entrance, park outside and climb over the wall.

Enter the Crypt

Now proceed to the entrance of the crypt, following the door icon.

Search the Tunnels

Jump down the hole in the ground and slide all the way down. When you reach the lanterns, you need to shoot their chains with a Batarang to then be able to access the ledges and continue on.

Keep going down the tunnel until a new objective unlocks.

Reach Other Side of the Collapsed Batcave

Grapple onto the lantern ledges to make it to the other side of the Batcave.

Search the Collapsed Batcave

Now squeeze through the cranny and resume sliding down, readying yourself to grapple onto a lantern ledge.

Then, drop down and defeat the Assassin. Now continue along the tunnels and down the slide and defeat the next three Assassins. Now continue past the gap and you'll have to defeat another Man-Bat.

Defeat the Man-Bat

As usual, avoid his swoop and wing attacks and only then can you land your standard or heavy melee attacks.

Search the Caverns

Now open the chest in the arena where you've fought the Man-Bat and continue past the joker card.

Squeeze through and keep running along the tunnels, then shoot your Batarang to release the chain from the lantern ledges and continue forward, sliding and grappling as you go.

Defeat the Assassins

At the bottom of the grappling and sliding sections, defeat the Assassins.

Search for an Exit

Open the chest in the area and then squeeze through the cranny.

Reach the Lower Caverns

Now continue down the slide and keep grappling your way forward.

Reach the Lower Caverns – Defeat the Assassins

At the bottom, defeat the enemies.

Reach the Lower Caverns

Squeeze through the crack and open the chest from the ledge halfway through.

Reach the Ancient Door

Drop down and defeat the Assassins, then shoot the chains to release the ledge and continue forward.

Now grab the chest on your left and continue forward.

Defeat the League Guards

Defeat the Assassins standing in your way. When you're done, open the chest by the entrance to the temple

Enter the Temple

Before squeezing through, make sure you have equipped your best gear, are powerful enough to face the two boss fights awaiting us and that you have enough Momentum Abilities at your disposal, especially the Tonfa Jackhammer.

Defeat Bruce Wayne

The entire fight is more of a matter of patience rather than skill. You basically need to avoid Bruce's attacks when they turn red and stepping on the Lazarus Pits in the arena. Make sure you keep avoiding and landing swift melee attacks to quickly recharge your Momentum. Once it's recharged, land Momentum Ability attacks to deal a bit more damage. Otherwise, take your time and dodge and you will eventually be able to plead to him for a third and final time.

Defeat Talia al Ghul

The fight against Talia spans over two phases and it's more or less the same as the one against Bruce, with the only difference that she lands much faster attacks, giving you very few windows of opportunity to land yours.

During phase one, she basically spams her attacks without interruption, and you need to try avoiding them at all times, leaving you very little time to land a few attacks, 1 or 2 at most. Over time your Momentum Abilities will charge up. When any Momentum Ability reaches its second bar, unleash it on Talia for some considerable damage.

During phase two, she uses the same fighting patterns, but this time wields a halberd and some of her attacks don't give you a red warning, especially after she rises in the air. Use the same strategy as in phase one and when two bars of Momentum Ability are charged up, land the attack for some considerable damage. Eventually, a cutscene will start, ending phase two.

The fight against Talia is the reason why we recommended making sure you're powerful enough before entering the temple. It's during this fight that you will need to be powerful enough to sustain lots of damage, but deal as much at the same time.

This finishes Case 08: Head of the Demon in Gotham Knights.

## 08: Head of the Demon

### 8.1 – Dangerous Skies

Check the Man-Bat Locations on the Evidence Boards

Approach the Evidence Boards and hover over the Leads section to complete this objective. Now we need to head outside and find three Man-Bats to defeat.

Defeat the Man-Bat 0/1

Head outside and proceed to the first Man-Bat's location in Elliot Center in the Financial District. The exact location of the Man-Bat is in the map screenshot below.

When fighting Man-Bats remember to avoid both his swoop and wing attacks. As you're fighting the Man-Bat, some League Assassins will show up as well.

Defeat the Man-Bat 0/1

Now proceed to the next Man-Bat's location on Gotham City General Hospital's rooftops in the West End district. The exact location of the Man-Bat is in the map screenshot below.

Defeat the Man-Bat 0/1

Now proceed to the third and final Man-Bat's location on WayneTech's rooftops in Southside. The exact location of the Man-Bat is in the map screenshot below.

# SUIT STYLES

This page contains information on all the different types of Suit Styles that can be equipped in Gotham Knights, including how to obtain them, the bonuses they contain, and how to swap out styles to keep the one you want.

In Gotham Knights, you'll be able to obtain various types of gear and items as you fight crime in Gotham, and will also learn the blueprints to creating an overall costume theme called a Suit Style. Regardless of the gear you have equipped, the Suit Style is a uniform that can vary in looks, and can also be tweaked to include certain differences - like revealing parts of the face or changing the emblem to be more prominent. The blueprints for these Suit Styles can often be found by defeating tough opponents like some of Gotham's Most Wanted, or undertaking other side quests. Like the gear you can equip, different Suit Styles will also give you new bonuses.

The list below contains all known Suit Styles in Gotham Knights. There will be 44 individual suit styles across all 4 characters, totalling to 11 individual suits per character. Note that while there are no microtransactions in Gotham Knights, the "Beyond" and "Knightwatch" Suits will be locked to the Deluxe and Collector's Edition at launch.

See the list below for the suit styles currently revealed, and check back at launch to find out what bonuses they offer, and where the blueprints for these styles can be found.

Stats: TBD

Unlocked: Default Suit Style

The Stock Suits are what each of the Gotham Knights will have equipped at the start of the game. They include a longer coat for Robin, an emblazoned chest piece for Nightwing, a purple and bright yellow outfit for Batgirl, and a brown chest gear for Red Hood.

*Beyond Suitstyle*

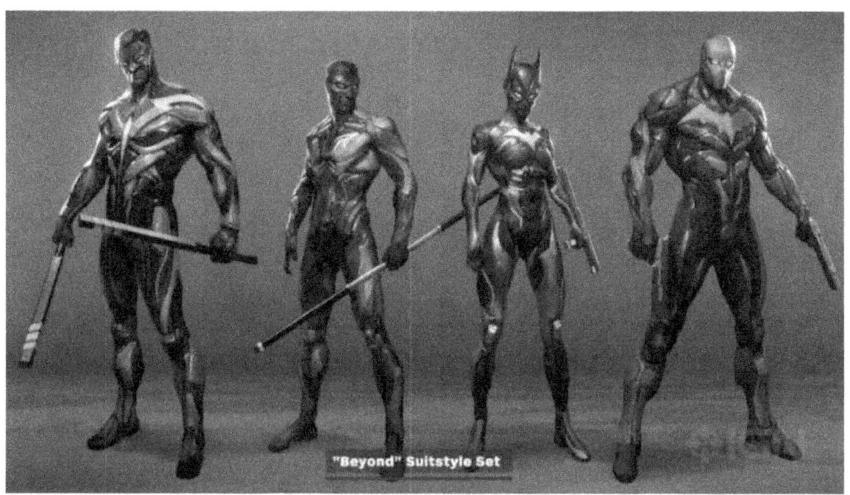


Stats: TBD

Unlocked: Currently locked to the Deluxe or Collector's Edition

The Beyond Suitstyle takes direct inspiration from the Batman Beyond animated series, as well as the Beyond Suit that was unlocked in the Arkham Knight game. It features a full face mask for each character, glowing eyes, and minimal stripes of color on black chrome.

## *Eternal Suitstyle*

Stats: TBD

Unlocks: TBD

The Eternal Suitstyle takes on a more comic-book style for each of the heroes, including tight suits with armored gloves and boots for most heroes, a more bulky metal defense for Red Hood, and also features Tim Drake's "Red Robin" inspired outfit.

### *Knight Ops Suitstyle*

Stats: TBD

Unlocks: TBD

The Knight Ops Suitstyle is a more urban combat theme for the Gotham Knights, featuring lots of pratical body armor, extra pouches and gear, and subdued colors for characters like Batgirl and Red Hood. It also features longer hair for Nightwing, and a trimmed hood and cape for Robin.

## *Knightwatch Suitstyle*

Stats: TDB

Unlocks: Currently locked to the Deluxe or Collector's Edition

Designed by famed comic book writer, illustrator, and current CCO of DC Comics Jim Lee. It features a very comic book-inspired set for the Gotham Knight that is very vibrant, including flashier colors, traditional outfits for Robin and Nightwing, Batgirl's classic black and gold color scheme, and a large brown jacket over Red Hood's armor.

## Shinobi Suitstyle

Stats: TBD

Unlocks: TBD

The Shinobi Suitstyle takes clear Japanese inspiration for its design, featuring kamishomo-styled upper clothing for Red Hood and Nightwing, a cobra-style hood for Robin, and a top-knot for Batgirl.

## *Year One Suitstyle*

Stats: TBD

Unlocks: TBD

The Year One Suitstyle are early and more informal urban clothing for the Gotham Knights, with less emphasis on suits, and more on altered jackets, pants, boots, and spare rugged body armor.

## Neon Noir

Stats: TBD

Unlocks: TBD

The Neon Noir Suitstyle features a more mature theme for the Knights. These suits are adapted to move fast in the city's shadows while creating an atmosphere of mystery and danger.

# CHARACTERS

## Batman (Bruce Wayne)

This page contains information on the character Batman (Bruce Wayne) and their biography in Gotham Knights. This includes their backstory in this particular setting of Gotham, what happened to them leading up to the events of the game, and how they impact the story.

### *Batman's Backstory in Gotham Knights*

Unlike the Batman in Rocksteady and WB Games' Batman Arkham series, this Batman did not undergo the events of Arkham Origins, Asylum, City, and Knight. He was able to build up and train his "Bat Family", starting with Dick Grayson as the original Robin, and then Jason Todd, Tim Drake, and Barbara Gordon.

It appears certain events that tie many Batman stories together have happened in this universe, including Jason Todd's supposed death at the hands of the Joker, turning him into the antihero Red Hood that still works with Batman on occassion. Batgirl was also paralzyed by the Joker for a time, but she has managed to recover and continue her crime fighting.

Gotham Knights appears to begin at the twilight of Batman's career, as they've no doubt encountered the usual villains time and time again - however a new force has been lurking unseen underneath Gotham for a long time, and is finally ready to show itself by striking at Batman first.

### *Batman's Role in Gotham Knights*

Gotham Knights begins with the death of Batman / Bruce Wayne. Unlike Rocksteady's Arkham Knight, this does not appear to be a planned move by Batman himself, but a strategic strike by someone with knowledge of his identity. In the event of his death, Batman triggered the Batcave to self-detonate, sending a "Code Black" to the rest of the Bat Family informing them of his passing, and to use the Belfry as a base of operations to continue their work.

While the culprit of Batman's assassination seems to point to the mysterious Court of Owls, the truth of his death remains to be seen over the course of Gotham Knights' story.

## Robin (Tim Drake)

This page contains information on the character Robin (Tim Drake) and their biography in Gotham Knights. This includes their backstory in this particular setting of Gotham, what happened to them leading up to the events of the game, and how they impact the story.

### Robin's Backstory in Gotham Knights

As the third crime fighter to take up the mantle of "Robin", Tim Drake was a genius and sleuth from a young age. The young prodigy was further trained by Batman and became convinced in his mission that Gotham City needs protecting. While he may be the smallest and youngest of the group, he's more than capable of handling himself using stealth, tech, and melee combat.

### Robin's Role in Gotham Knights

As one of four playable characters in Gotham Knighs, Robin is an expert fighter armed with his collapsible quarterstaff and skilled in a variety of stealth techniques, Robin also possesses a background in combined psychological warfare and behavioral sciences, and his skillset includes using a variety of advanced tech to counter his smaller size.

In addition to his staff that can be used to deflect bullets, Robin also has a high powered slingshot full of different types of projectiles to distract and disrupt, is able to utilize cloaking technology, and can even use the Justice League's satellite to teleport short distances to get around behind enemies.

# Nightwing (Dick Grayson)

This page contains information on the character Nightwing (Dick Grayson) and their biography in Gotham Knights. This includes their backstory in this particular setting of Gotham, what happened to them leading up to the events of the game, and how they impact the story.

## *Nightwing's Backstory in Gotham Knights*

The original sidekick to use the name "Robin", Dick Grayson was Batman's first protege, and is a master in acrobatics thanks to his upbringing in the circus before his parents were gunned down - echoing Bruce Wayne's own tragedy. After saving Gotham together with Batman for years, he eventually left to become a hero in his own right under the name Nightwing. Leaving Gotham for the distant crime-ridden city of Bludhaven, Nightwing remained in close communication with Batman to help when needed, and now Gotham will need the original boy wonder more than ever.

## *Nightwing's Role in Gotham Knights*

As one of four playable characters in Gotham Knights, Nightwing utlizies superior movement and acrobactic combat to dispatch foes with precision. Using his twin sticks, Nightwing can quickly take out any threat in range, and can even take out targets at ranged with wrist projectiles, electric attacks, and his ricochet moves.

Able to use both his standard grappler and a winged glider to traverse great distances, no foe can stay out of Nightwing's reach for long.

## Batgirl (Barbara Gordon)

This page contains information on the character Batgirl (Barbara Gordon) and their biography in Gotham Knights. This includes their backstory in this particular setting of Gotham, what happened to them leading up to the events of the game, and how they impact the story.

### *Batgirl's Backstory in Gotham Knights*

The daughter of the late Commissioner Gordon of the Gotham City Police Department, Barbara Gordon was obsessed with fighting crime from a young age. Rather than follow her father into law enforcement, Barbara sought to follow Batman's example of crime fighting, and took on the name Batgirl as she was trained by the caped crusader.

A devastating attack by The Joker left Batgirl unable to fight crime physically, and so she became known as The Oracle for a time to continue aiding her friends. But with surgery, physical therapy, and time, she is finally back and ready to fight crime once more as a superhero.

### *Batgirl's Role in Gotham Knights*

As one of four playable characters in Gotham Knights, Batgirl wields a tonfa that can change its shape to suit her style, attacking fast or hitting hard interchangeably. She possesses a wealth of different gadgets - ranging from a storm of batarangs to a claw that can pull herself (or her enemies) into striking distance.

To traverse Gotham quickly, she can glide far distances with her cape, or race down the streets in her stylish motorcycle to track down criminals.

## Red Hood (Jason Todd)

This page contains information on the character Red Hood (Jason Todd) and their biography in Gotham Knights. This includes their backstory in this particular setting of Gotham, what happened to them leading up to the events of the game, and how they impact the story.

### Red Hood (Jason Todd)'s Backstory in Gotham Knights

The second crime fighter that took on the mantle of "Robin", Jason Todd was a streetwise vigilante that sought out Batman when they were young to learn the tricks of the trade. Tragically, their brash nature led them right into a trap - with the Joker brutally killing Jason Todd. He was then resurrected by the Lazarus Pit, and their rage twisted them into a brutal killer who came back to Gotham much later as the Red Hood.

While they've since patched things up with Batman and the rest of the family (and now use non-lethal rounds for their guns), Red Hood still prefers a more explosive approach to crime fighting.

### Red Hood (Jason Todd)'s Role in Gotham Knights

As one of four playable characters in Gotham Knights, Red Hood boasts extreme ranged firepower in the form of their twin pistols that they can use to deal incredible damage. They've also begun to embrace more mystical arts stemming from their unnatural resurrection. This allows them to perform jumps in tandem - using the very air as platforms to reach new heights without the aid of gadgets or gliders.

# Harley Quinn

This page contains information on the character Harley Quinn(Dr. Harleen Quinzel) and her biography in Gotham Knights. Here you'll find her backstory in this particular setting of Gotham, what happened to her leading up to the game's events, and how those events impact the story.

## *Harley's Backstory in Gotham Knights*

Harley started as a psychiatric intern in Arkham Asylum when she was still going by the name Dr. Harleen Frances Quinzel. After finally being able to treat The Joker, he seduced her, resulting in her falling madly in love with him, leading her to turn into Harley Quinn, Joker's sidekick.

Their relationship was incredibly complex and toxic, leading to the Joker attempting to kill her more than once. These circumstances prompted Harley straight to the arms of Poison Ivy, who would end up being Quinn's second love interest.

Eventually, Harley would even collaborate with the Dark Knight himself to bring Joker down, and she would also join the Suicide Squad. In Gotham Knights, players will encounter a Harley Quinn who has already been through all of what we discussed before, so she is older, wiser, and no longer held by other people's whims.

### *Harley's Role in Gotham Knights*

In Gotham Knights, Harley is no longer the sidekick who follows Joker around like a lost puppy or someone who is learning to walk the fine line between being evil and doing things for the greater good (like she is when she joins the Suicide Squad). It's important to know that Gotham Knight's Harley isn't a different version of the character; she's just a Harley that's further down her timeline, so it's a Harley players are not used to seeing.

## Clayface

### *Clayface's Backstory in Gotham Knights*

As Basil Karlo, Clayface was an actor who went mad after learning that a classic film where he was the protagonist would be remade with a different actor as the lead. This descent into madness prompted him to kill several of the new movie's cast members before batman stopped him.

He then joined a group of mutated villains called the Mud Pack and tricked them into gaining their power. This event led Karlo to become the giant mud monster we will see in Gotham Knights.

### Clayface's Role in Gotham Knights

There isn't much information about Clayface and his role in Gotham Knights. We know he will appear in his mud form and that whatever he plans has to do with a big Premier.

As for his powers, he can go through bars (as seen in the sewers scene), morph his body to grow extra limbs, and he probably isn't working alone, as we can see a clay army running through the sewers in one of the game's trailers.

## Mister Freeze

### Mister Freeze's Backstory in Gotham Knights

Mister Freeze is a cryogenics expert whose life changed forever when he met the love of his life   — Nora.

Although his backstory details change from version to version, what remains the same is that an icy explosion occurred in his lab while he was trying to save Nora.

Somehow Mister Freeze made it out alive, but now he could only survive in subzero temperatures, which prompted him to construct a suit that could maintain these demands. To support this new lifestyle, he started robbing diamonds to power up his suit and get money to fund his experiments.

### *Mister Freeze's Role in Gotham Knights*

In Gotham Knights, Mister Freeze sports his usual Freeze Gun, a technologically advanced weapon that can freeze enemies. Not only that, but he can also deal damage using ice-enhanced elemental attacks.

As for his motivations, all we know is that Freeze is back in Gotham with a mysterious plan to change the city's weather so "the world will feel the cold as he does."

But Mister Freeze won't be working alone in Gotham Knights. For a portion of his story, you will see him teaming up with The Regulators — a brand new faction in the city that likes to steal advanced goods, traffic organs, and increase their own powers using tech.

## Court of Owls

This page contains information on the secret organization known as the Court of Owls and their role in Gotham Knights. This includes their backstory in this particular setting of Gotham, what happened to them leading up to the events of the game, and how they impact the story.

### *Court of Owls's Backstory in Gotham Knights*

The Court of Owls is a secret society that has worked in the shadows to control Gotham City for a very long time. The members of the Court of Owls are part of the oldest and wealthiest families in Gotham, so all this time, they have been shaping the city's history and controlling the flow of wealth for more than 200 years.

### *Court of Owls' Role in Gotham Knights*

The Court of Owls are set to big the big villains in Gotham Knights. The problem is that most of Gotham's inhabitants think of them as a work of fiction, a folktale, rather than a real threat. Even the Knights don't believe they're real, as we can see Nightwing saying in one of the trailers where he speaks to the Penguin.

But the Court is real, and they're everywhere, always listening and plotting in the shadows. Keep in mind that the Court's power is not only in their secrecy and money; they also have hired assassins, known as Talons, that do their dirty work for them and will no doubt represent a significant obstacle when exploring Gotham.

# The Penguin

This page contains information on the character The Penguin (Oswald Cobblepot) and his biography in Gotham Knights. This includes his backstory, what happened to him leading up to the game's events, and how they impact the story.

### The Penguin's Backstory in Gotham Knights

The Penguin is one of Batman's emblematic enemies, almost as legendary as the Joker himself.

The Penguin is a crime lord at heart, so most of his evil doings revolve around illegal financial activities that fund most of Gotham's organized crime.

### The Penguin's Role in Gotham Knights

We know that the Penguin will appear in Gotham Knight to warn the Knights about looking further into The Court of Owls organization. However, it is not yet confirmed if he will be a prominent villain like Harley or Mister Freeze are going to be.

# SKILL TREES

This page contains information about the different Skill Trees available in Gotham Knights. Here you'll find what each character's Skill Trees focus on and what you will get once you unlock the Knighthood Skill Tree. A core mechanic in Gotham Knights is the ability to play as four characters with different playstyles, so you can choose whatever feels best for you. Each character has four skill trees that allow you to explore different directions to specialize your main.

Looking for something in particular? Click the links below to jump to...

## How to Unlock Skill Trees

There seem to be seven abilities per tree; you can unlock them by spending Ability Points on them. You will gain these points just by playing the game.

Something to note is that when you get Ability Points for your main character, you will have a surplus of points for the others, allowing you to upgrade their Skill Trees without having to replay certain game sections. This way, you won't feel like you're being penalized for changing characters mid-playthrough.

## How Skill Trees Work

There are four different skill trees for each character. Three of them are unique for that character, while the fourth (called Knighthood) is available to all of them.

## What is The Knighthood Skill Tree

The Knighthood Skill Tree represents the moment in the storyline when your character finally realizes they can be the Dark Knight

Gotham needs, rather than just a new Batman. For Robin, this moment comes when he realizes he can tap into the Justice League Satellite and use it to fight crime.

This Skill Tree will unlock after a specific point in the story that you won't need to play again whenever you want to unlock the Knighthood tree. However, you will still need to complete a set of unique challenges for each character to unlock their Knighthood abilities.

Unlocking the Knighthood Skill Tree not only unlocks the abilities within this tree but also unlocks each character's traversal method and an ultimate ability.

## Red Hood Skill Trees

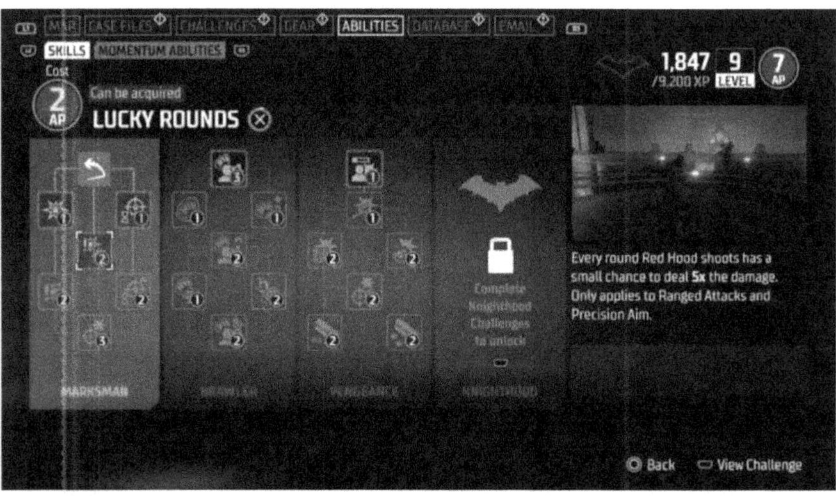


Red Hood specializes in ranged combat, which shows up in his Skill Trees. For example, you can unlock:

### *The Marksman Skill Tree*

The abilities in this tree are all about dealing the most damage per second, so you'll be rewarded for shooting and precision.

## All Abilities in the Marksman Skill Tree

| Icon | Ability | What it does | Requirement | Cost |
|---|---|---|---|---|
|  | Perfect Evade | Red Hood performs a perfectly-timed evade that generates Momentum and allows for a perfect attack follow-up | Unlock the Marksman Skill Tree | TBA |
|  | Focused Fire | Red Hood can aim longer at a target to deal 4x more damage | Perfect Evade | 1 Action Point |
|  | Lucky Rounds | Every round Red Hood shoots has a small chance to deal 5x the damage. Only applies to Ranged Attacks and Precision Aim | Perfect Evade | 2 Action Points |
|  | Critical Expertise | Increases Red Hood's critical damage by 20% | Perfect Evade | 1 Action Point |

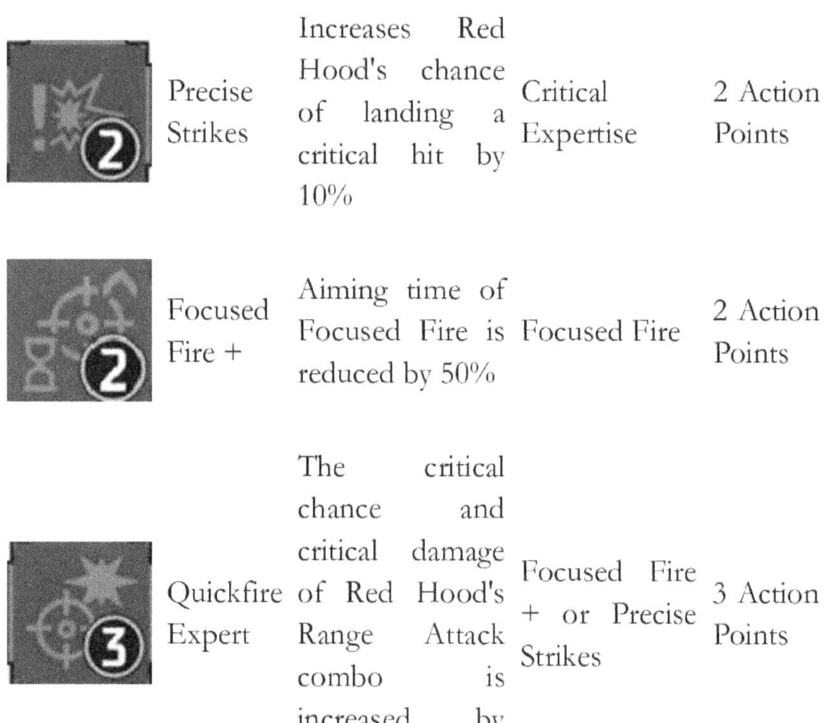

| Icon | Ability | What it does | Requirement | Cost |
|---|---|---|---|---|
| | Precise Strikes | Increases Red Hood's chance of landing a critical hit by 10% | Critical Expertise | 2 Action Points |
| | Focused Fire + | Aiming time of Focused Fire is reduced by 50% | Focused Fire | 2 Action Points |
| | Quickfire Expert | The critical chance and critical damage of Red Hood's Range Attack combo is increased by 15% | Focused Fire + or Precise Strikes | 3 Action Points |



### *The Brawler Skill Tree*

Brawler abilities are all about melee combat and grab and throw mechanics. The abilities in this tree focus on the fact the Red Hood is the biggest character out of the four Knights.

All Abilities in the Brawler Skill Tree

| Icon | Ability | What it does | Requirement | Cost |
|---|---|---|---|---|
| | Human Bomb | When Red Hood throws an enemy, he attaches a concussion mine to | Unlock the Brawler Skill Tree | 3 Action Points |

| Icon | Name | Description | Requires | Cost |
|---|---|---|---|---|
| | | them that explodes when shot | | |
| | Large Grab | Red Hood can perform grab moves on large enemies | Human Bomb | 1 Action Point |
| | Extended Grab Window | Enemies can now be grabbed at 50% health or less | Human Bomb | 1 Action Point |
| | Human Bomb Enhanced | Increases damage and radius of the concussion mine explosion | Large Grab or Extended Grab Window | 2 Action Points |
| | Grip Expertise | Increases Red Hood's damage when performing a grab move by 10% | Human Bomb Enhanced | 1 Action Point |
| | Iron Grip | Grabbing an enemy prevents Red Hood from being interrupted by most attacks | Human Bomb Enhanced | 2 Action Points |
| | Human Bomb Multiplied | Detonating the concussion mine deploys 6 additional concussion proximity mines on | Iron Grip or Grip Expertise | 2 Action Points |

the ground

### The Vengeance Skill Tree

Red Hood is not only a big man but also filled with rage. This strong emotion is what the Vengeance Skill Tree focuses on, so you'll see in this tree abilities that deal more damage to certain enemies types that he is especially angry towards.

All Abilities in the Vengeance Skill Tree

| Icon | Ability | What it does | Requirement | Cost |
|---|---|---|---|---|
| | Coup De Grace | Red Hood inflicts 10% more damage on enemies with 30% health or less | Unlock the Vengeance Skill Tree | 1 Action Point |
| | Freak Justice | Increases Red Hood's damage by 15% and critical damage by 5% when fighting Freaks | Coup De Grace | 1 Action Point |
| | Regulator Justice | Increases Red Hood's damage by 15% and critical damage by 5% when fighting the Regulators | Freak Justice | 2 Action Points |

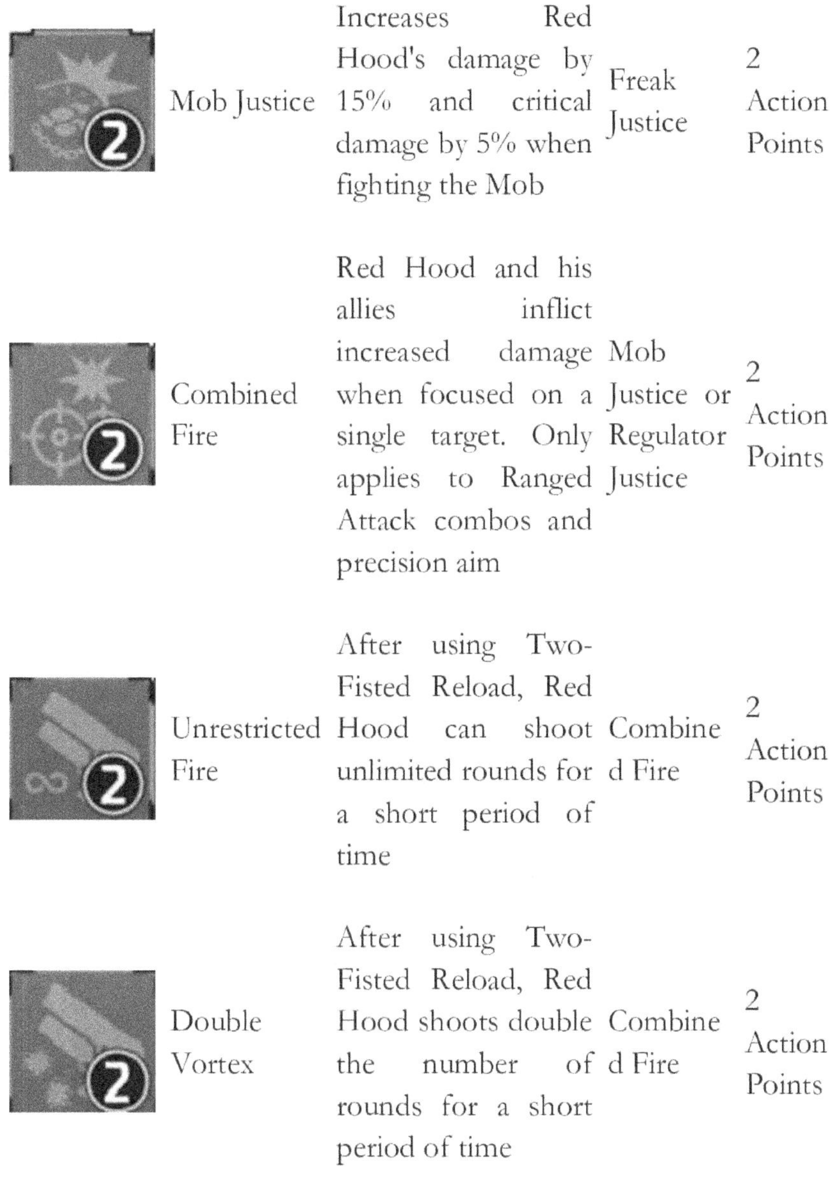

| | | | | |
|---|---|---|---|---|
| | Mob Justice | Increases Red Hood's damage by 15% and critical damage by 5% when fighting the Mob | Freak Justice | 2 Action Points |
| | Combined Fire | Red Hood and his allies inflict increased damage when focused on a single target. Only applies to Ranged Attack combos and precision aim | Mob Justice or Regulator Justice | 2 Action Points |
| | Unrestricted Fire | After using Two-Fisted Reload, Red Hood can shoot unlimited rounds for a short period of time | Combined Fire | 2 Action Points |
| | Double Vortex | After using Two-Fisted Reload, Red Hood shoots double the number of rounds for a short period of time | Combined Fire | 2 Action Points |

## Red Hood's Knighthood Skill Tree

When you unlock this tree for Red Good, you'll also be able to use a special ability called "Mystical Leap" to move around the city using spirit platforms.

All Abilities in Red Hood's Knighthood Skill Tree

| Icon | Ability | What it does | Cost |
| --- | --- | --- | --- |
|  | Mystical Leap | Red Hood traverses through the air using spirit platforms | TBA |
|  | Ranged Terror | Every shot that knocks out a target inflicts Fear in nearby enemies | 1 Action Point |
|  | Weak Spot Damage+ | Increases headshot and weak spot damage by 15% | 1 Action Point |
|  | Grab Dread | Grabbing a target inflicts Fear in nearby enemies | 1 Action Point |
|  | Combat Mastery | Increases the number of attacks in Red Hood's Ranged Attack combo by 1. The last hit is a knockdown | TBA |

 Ducra's Training | Mystical Rounds requires 50% less time to lock on targets | TBA

 Shadow Vengeance | Mystical Rounds shoots 2 rounds instead of 1 | 3 Action Points

## Batgirl Skill Trees

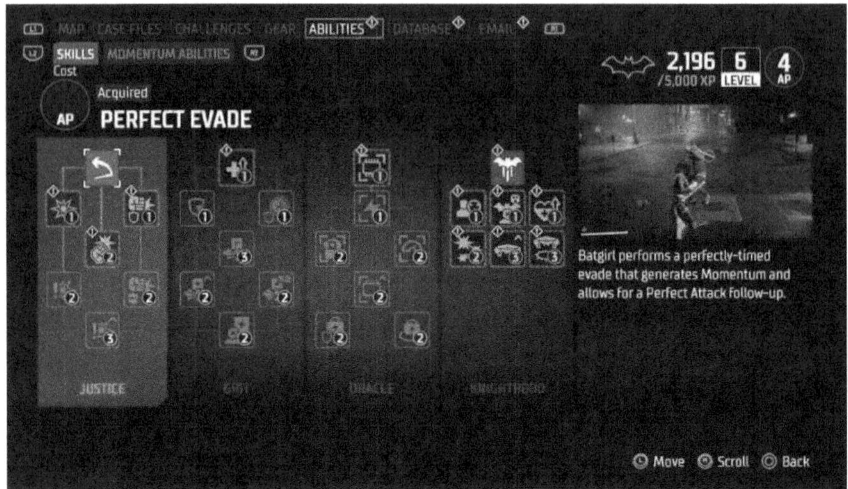


Batgirl specializes in highly focused single-target combat, which shows up in her Skill Trees. For example, you can unlock:

### The Justice Skill Tree

The abilities in this tree are all about dealing the more critical hits with more power, and increasing Beatdown damage on single targets.

All Abilities in the Justice Skill Tree

| Icon | Ability | What it does | Requirement | Cost |
|---|---|---|---|---|
| | Perfect Evade | Batgirl performs a perfectly-timed evade that generates Momentum and allows for a perfect attack | Unlock the Justice Skill Tree | TBA |

follow-up

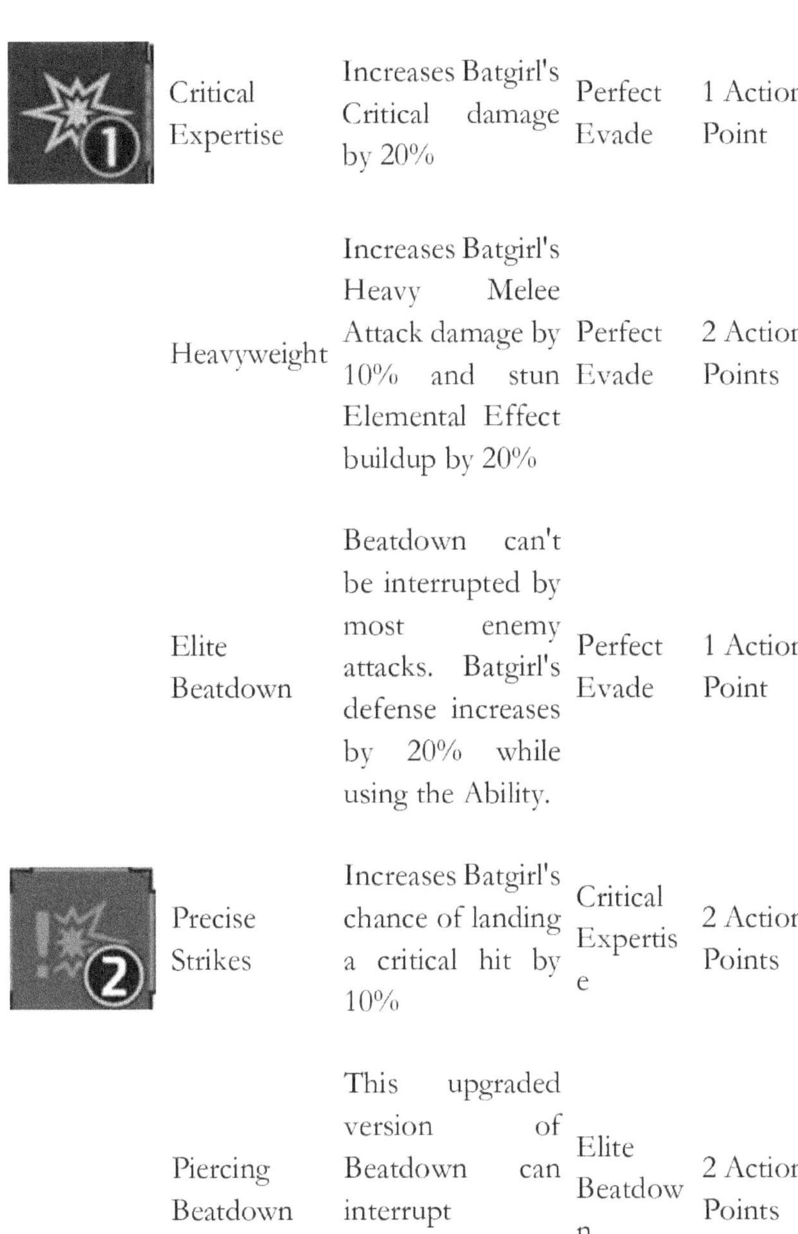

| | | | | |
|---|---|---|---|---|
| | Critical Expertise | Increases Batgirl's Critical damage by 20% | Perfect Evade | 1 Action Point |
| | Heavyweight | Increases Batgirl's Heavy Melee Attack damage by 10% and stun Elemental Effect buildup by 20% | Perfect Evade | 2 Action Points |
| | Elite Beatdown | Beatdown can't be interrupted by most enemy attacks. Batgirl's defense increases by 20% while using the Ability. | Perfect Evade | 1 Action Point |
| | Precise Strikes | Increases Batgirl's chance of landing a critical hit by 10% | Critical Expertise | 2 Action Points |
| | Piercing Beatdown | This upgraded version of Beatdown can interrupt enemies' Armored Attacks | Elite Beatdown | 2 Action Points |

| Critical Focus | Increases Batgirl's chance of landing a Critical hit by 10%. Increases Precise Critical damage by 10%. | Focused Fire + Precise Strikes | 3 Action Points |

## The Grit Skill Tree

The Grit tree appears to focus mainly on Batgirl being able to power through attacks and fight overwhelming odds by bolstering her defensive abilities to stay in the fight longer.

All Abilities in the Grit Skill Tree

| Icon | Ability | What it does | Requirement | Cost |
|---|---|---|---|---|
| | HP Plus | Increases Batgirl's base health by 40%. Does not apply to bonus health provided by gear. | Unlock the Grit Skill Tree | 1 Action Point |
| | TBA | | | |
| | TBA | | | |
| | TBA | | | |
| | TBA | | | |
| | TBA | | | |
| | TBA | | | |

| The Oracle Skill Tree

Using her knowledge from her time as the tech expert Oracle, this skill tree utilizes her ability to summon drones and hack the environment to get the drop on enemies.

All Abilities in the Oracle Skill Tree

| Icon | Ability | What it does | Requirement | Cost |
| --- | --- | --- | --- | --- |
| | Remote Hacking | Using AR Mode, Batgirl can disable certain devices, like cameras, turrets, mines, electronic panels, and laser control modules. Tap the prompt while aiming at a hackable electronic device. | Unlock the Oracle Skill Tree | 1 Action Point |
| | TBA | | | |
| | TBA | | | |
| | TBA | | | |
| | TBA | | | |
| | TBA | | | |
| | TBA | | | |

### *Batgirl's Knighthood Skill Tree*

Unlocking this tree for Batgirl will also give you the ability to glide and her ultimate ability, which is a drone.

## Robin Skill Trees

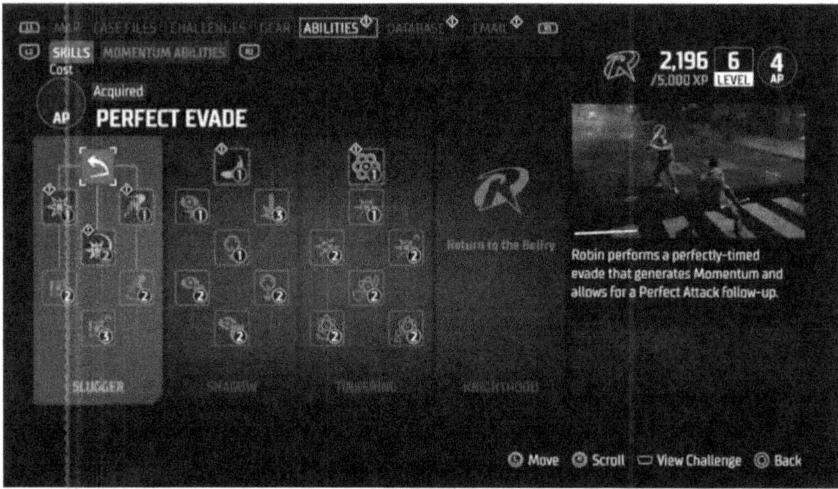

Robin specializes in stealth, subterfuge, and distractions, which shows up in his Skill Trees. For example, you can unlock:

### *The Slugger Skill Tree*

The abilities in this tree are all about dealing the more critical hits with more power, and inflicting Elemental Effects with the help of Robin's decoy.

All Abilities in the Slugger Skill Tree

| Icon | Ability | What it does | Requirement | Cost |
|------|---------|--------------|-------------|------|

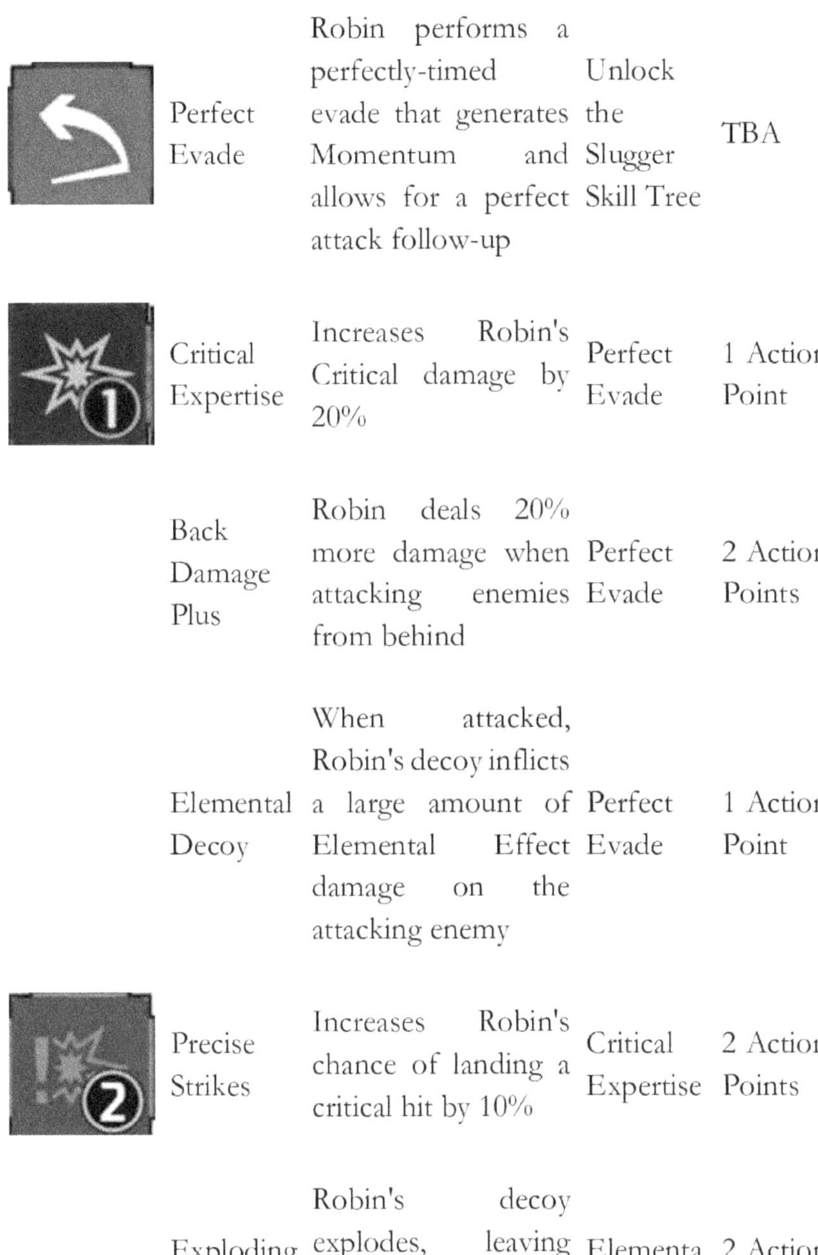

| | | | | |
|---|---|---|---|---|
| | Perfect Evade | Robin performs a perfectly-timed evade that generates Momentum and allows for a perfect attack follow-up | Unlock the Slugger Skill Tree | TBA |
| | Critical Expertise | Increases Robin's Critical damage by 20% | Perfect Evade | 1 Action Point |
| | Back Damage Plus | Robin deals 20% more damage when attacking enemies from behind | Perfect Evade | 2 Action Points |
| | Elemental Decoy | When attacked, Robin's decoy inflicts a large amount of Elemental Effect damage on the attacking enemy | Perfect Evade | 1 Action Point |
| | Precise Strikes | Increases Robin's chance of landing a critical hit by 10% | Critical Expertise | 2 Action Points |
| | Exploding Decoy | Robin's decoy explodes, leaving behind Elemental residue that deals Elemental Effect | Elemental Decoy | 2 Action Points |

| | Ability | What it does | Requirement | Cost |
|---|---|---|---|---|
| | | damage to any enemy that walks into it | | |
| | Elemental Focus | Robin's Critical damage and chance of landing a Critical Hit increase by 20% on targets with an active Elemental Effect | Exploding Decoy or Precise Strikes | 3 Action Points |

## The Shadow Skill Tree

The Shadow Skill tree fully focuses on Robin's stealth abilities, allowing him to move undetected, strike harder from stealth, and perform Batman's signature vantage point takedowns.

All Abilities in the Shadow Skill Tree

| Icon | Ability | What it does | Requirement | Cost |
|---|---|---|---|---|
| | Light Footed | Robins produces no sound while running and moves faster while crouched (Does not apply to while sprinting) | Unlock the Shadow Skill Tree | 1 Action Point |
| | Stealth Damage | Robin's damage is increased while | Shadow Renewal | 1 Action Point |

| | | | |
|---|---|---|---|
| Plus | undetected. Melee and Ranged attacks by +10%. Takedowns and Stealth Strikes by 20% | | |
| Turnabout Takedown | Robin can perform Takedowns and Stealth Strikes on large enemies | Shadow Renewal | 3 Action Points |
| Vantage Hanging Takedown | Robin can perform a special takedown that suspends enemies from a vantage point. While perched on a vantage point, tap the prompt when an enemy is below Robin | Shadow Renewal Stealth Damage Plus or Turnabout Takedown | 1 Action Point |
| Reduced Visual Stim | Robin is harder to spot and enemies take longer to notice him | Vantage Hanging Takedown | 2 Action Points |
| Vantage Hanging Takedown Mine | Performing a Vantage Hanging Takedown also drops a proximity mine that lures enemies. | Vantage Hanging Takedown | 2 Action Points |
| Shadow Renewal | Successful takedowns restore 25% of | Reduced Visual Stim or Vantage | 2 Action Points |

|  |  |
| --- | --- |
| Robin's health | Hanging Takedown Mine |

## The Tinkering Skill Tree

The Tinkering tree lets Robin maximize his elemental damage inflicted upon enemies, and also improves his ranged combat to place elemental mines for approaching thugs.

All Abilities in the Tinkering Skill Tree

| Icon | Ability | What it does | Requirement | Cost |
| --- | --- | --- | --- | --- |
|  | Elemental Charge | Elemental Effect build up inflicted by Robin is increased 25% faster | Unlock the Tinkering Skill Tree | 1 Action Point |
|  | Sticky Pellet | In Precision Aim, Robin's pellet can stick to surfaces, creating a temporary mine. Placing a second mine deactivates the first | Elemental Charge | 1 Action Point |
|  | Sticky Pellets x3 | Robin can place up to 3 sticky pellets at once | Sticky Pellet | 2 Action Points |
|  | Enhanced Sticky Pellets | Sticky pellets inflict 33% more damage and last up to 10 seconds | Sticky Pellet | 2 Action Points |

| Elemental Resistance | Robin's Elemental Effect resistance increases by 40% | Sticky Pellets x3 or Enhanced Sticky Pellets | 2 Action Points |
| --- | --- | --- | --- |
| Elemental Burst | When Robin deals Elemental Damage, he gains a 5% chance of dealing additional Elemental Damage in an area. Elemental Burst also reduces enemies' Elemental Effect resistance by 50% for 10 seconds. | Elemental Resistance | 2 Action Points |
| Enhanced Elemental Effects | Elemental Effects inflicted by Robin last twice as long | Elemental Resistance | 2 Action Points |

## Robin's Knighthood Skill Tree

When you unlock this tree for Robin, you'll also get the ability to do short-range teleportation.

# Nightwing Skill Trees

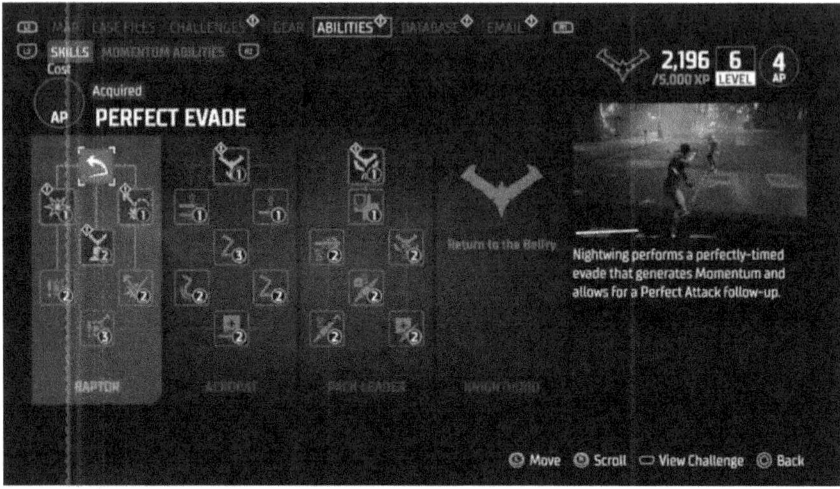

Nightwing specializes in mobility and group combat, as well as rallying the team to even greater heights.

## The Raptor Skill Tree

The abilities in this tree are all about dealing the more critical hits with more power, and increasing Beatdown damage on single targets.

All Abilities in the Raptor Skill Tree

| Icon | Ability | What it does | Requirement | Cost |
|---|---|---|---|---|
|  | Perfect Evade | Nightwing performs a perfectly-timed evade that generates Momentum and allows for a perfect attack | Unlock the Raptor Skill Tree | TBA |

| | | | | |
|---|---|---|---|---|
| | | follow-up | | |
|   | Critical Expertise | Increases Nighwing's Critical damage by 20% | Perfect Evade | 1 Action Point |
| | Assassin's Mark | Marks an enemy, increasing Nightwing and his allies' damage to the foe by 10%. Hold the prompt while aiming at an enemy | Perfect Evade | 2 Action Points |
| | Trampoline | Nightwing's Pounce ability is automatically followed by a high jump on the enemy. This ability cannot be used if Nightwing is under a low ceiling | Perfect Evade | 1 Action Point |
| | Precise Strikes | Increases Nightwing's chance of landing a critical | Critical Expertise | 2 Action Points |

| Icon | Ability | What it does | Requirement | Cost |
|---|---|---|---|---|
| | | hit by 10% | | |
| | Aerial Bounce | Nightwing bounces off an enemy following an aerial attack to propel himself back into the air. Can be used up to 3 times in succession | Trampoline | 2 Action Points |
| | Critical Distance | Hitting an enemy with a Melee Attack from a large distance increases the Critical chance and Critical Damage by 15% | Aerial Bounce or Precise Strikes | 3 Action Points |

## The Acrobat Skill Tree

The Acrobat tree focuses on Nightwing's ability to gain momentum from a variety of evasive actions and bolster his abilities from using Momentum Abilities.

All Abilities in the Acrobat Skill Tree

| Icon | Ability | What it does | Requirement | Cost |
|---|---|---|---|---|
| | Aerial | Increases Nightwing's | Unlock the | 1 Action |

| | | | |
|---|---|---|---|
| Damage Plus | aerial attack damage by 20% | Acrobat Skill Tree | Point |
| Extra Momentum Bar | Grants Nightwing an extra Momentum bar | Aerial Damage Plus | 1 Action Point |
| Momentum Gain Plus | Increases Nightwing's Momentum gain by 15% | Aerial Damage Plus | 1 Action Point |
| Evade Chain | Nightwing chain evades by performing a quick succession of back jumps | Extra Momentum Bar or Momentum Gain Plus | 3 Action Points |
| Haly's Favorite | Completing an Evade Chain knocks down all nearby enemies | Evade Chain | 2 Action Points |
| Evade Chain Momentum | Performing an evade chain during combat restores a portion of Nightwing's Momentum | Evade Chain | 2 Action Bars |
| Mind and Body | Using Momentum Abilities restores a portion of Nightwing's Health | Haly's Favorite or Evade Chain Momentum | 2 Action Bars |

## The Pack Lader Skill Tree

As the name implies, the Pack Leader skill tree is best used in multiplayer, as most of its skills rely on having a teammate to bolster their abilities or heal them in battle.

All Abilities in the Pack Leader Skill Tree

| Icon | Ability | What it does | Requirement | Cost |
| --- | --- | --- | --- | --- |
| | Family Ties | Increases Nightwing's defense and resistance by 10%. Working with allies also grants him additional bonuses: Batgirl - Melee Damage +15%. Red Hood - Ranged Damage +15%. Robin - Stealth Damage +15% | Unlock the Pack Leader Skill Tree | 1 Action Point |
| | Health Bolstered Defense | When Nightwing's Health is at least 70% or more, he gets a +5% Defense bonus scaling up to 20% when at full health. | Family Ties | 1 Action Point |
| | Momentum Regen | Nightwing's Momentum regenerates over time. Regeneration stops after filling 1 | Health Bolstered Defense | 2 Action Points |

| | | | |
|---|---|---|---|
| | Momentum segment. Gaining any Momentum activates regeneration again. Working with allies increases regeneration spee | | |
| Shared Skill | Passive Skills increasing Damage, Critical Chance, Defense, Momentum Regeneration, and Ultimate Cooldown are shared with allies at 50% of their value | Health Bolstered Defense | 2 Action Points |
| Elemental Smart Darts | Nightwing's darts inflict Elemental Effect build up on enemies or heal allies over time. | Momentum Regen or Shared Skill | 2 Action Points |
| Elemental Smart Darts Plus | Nightwing's darts also reduce enemies' defense by 10% and increase damage inflicted by allies by 5% for 10 seconds | Elemental Smart Darts | 2 Action Points |
| Revive Darts | Nightwing's darts can be used to instantly revive an ally from | Elemental Smart Darts | 2 Action Points |

afar. Maximum 1 use
per night

### *Nightwing's Knighthood Skill Tree*

When you unlock this tree for Nightwing, you'll also be able to use his glider, the Flying Trapeze, to get around the city.

## What are Momentum Abilities

Just like Skill Trees, Momentum Abilities are another set of unlockables that can change your character's playstyle and deal more damage.

These second sets of abilities are ones that you assign to specific buttons so you can achieve a series of cool-looking attack combos.

A cool thing about Momentum Abilities is that they pair well with abilities gained through Skill Trees. For example, a Momentum ability — like Batgirl's beat down — can be upgraded by unlocking other abilities in her Skill Trees so she can perform the beat down more effectively and with piercing damage.

## How to Unlock Momentum Abilities

To unlock these abilities, you must engage in different challenges throughout Gotham.

### *Red Hood's Momentum Abilities*

Below you'll find all the Momentum Abilities available for Red Hood.

All of Red Hood's Momentum Abilities

| Icon | Momentum Ability | What it does | Momentum Cost |
| --- | --- | --- | --- |
| | Barrage | Red Hood Shoots multiple rounds straight ahead, hitting all enemies in his path. Ammo is also reloaded. | 1 |
| | Mystical Punch | Red Hood charges his fists with Elemental Effect damage. Melee Attacks do 50% more damage and 5x more Elemental buildup. Duration: 30 s | 1 |
| | Two-Fisted Reload | Red Hood Performs a quick reload while hitting and shooting at an enemy | 1 |
| | Harnessed Rage | When activated, Red Hood's next attack increases by a portion of damage he receives. Also heals Red Hood at 1% per second. Duration: 20 s | 1 |

**Spoilsport Reload** — Red Hood jumps backward, releasing explosive magazines. Ammo is also reloaded. — 2

**Portable Turret** — Red Hood deploys a mini turret that shoots at nearby enemies, dealing high damage.
Duration: 6 s — 2

**Mega Tackle** — Red Hood dashes forward, knocking down all enemies in his path — 2

**Mystical Rounds** — Red Hood Locks onto enemies and shoots a Mystical Round that will hit all targets, causing massive damage — TBA

# CRAFTING GUIDE

This page contains information about the different Crafting Systems available in Gotham Knights. Here you'll find how the crafting system works in this game and what kind of equipment you will be able to craft.

## How Crafting Works in Gotham Knights

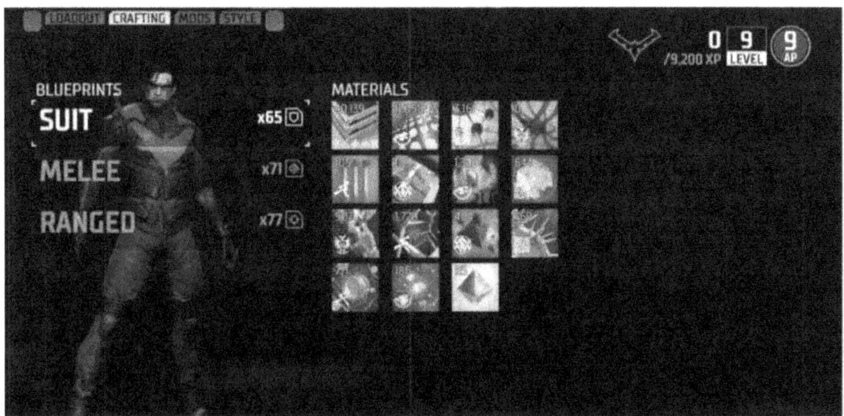

During the official Nightwing and Red Hood gameplay demo, we saw how the crafting system works in Gotham Knights. To upgrade the equipment you're working with, go to the workbench in the Belfry and interact with it.

In the crafting menu, you'll see your available blueprints and the materials you need to craft them. You'll be able to find said blueprints and crafting materials by taking on missions and fighting crime in Gotham.

## What Can You Craft in Gotham Knights

From the time being, it looks like you'll be able to craft suits and melee and ranged weapons.

## Suit Crafting

You'll be able to craft a wide array of suits to find the best fit for the challenges ahead. Below is a list of all the stats each suit in the game will have:

All Suit Stats in Gotham Knights

| Stat | What it does |
| --- | --- |
| Rarity | this one indicates the suit's quality. The rarer the suit, the better it'll be. |
| Level | It's unclear what this stat refers to, but it could indicate the level your character has to be to equip the suit. |
| Grade | A stat that's only available for Legendary suits. |
| Power | A suit's power level most likely indicates how much extra damage your character can deal when equipping said suit. |
| Armor | This one indicates how much damage the suit negates. |
| Health | This stats lets you know how many hits you will be able to endure while wearing a suit. |
| Momentum Generation | Not all suits have this stat, but it looks like it's tied to the ability you'll have to charge up before executing a specific move. |
| Elemental | Suits with this stat will be able to negate specific |

Armor      and  elemental attacks.
Resistance

Besides the stats mentioned above, some suits will have Mod Slots so you can equip mods and further improve the suit's abilities.

Finally, suits also have a particular Style, which we discussed more in-depth in our Suit Styles guide.

### Weapon Crafting

From what we've seen so far, you'll be able to craft both Melee and Ranged weapons. Just like suits, these weapons will have different stats, including elemental stats, so you'll be able to deal elemental damage with some of them.

# WHICH CHARACTER SHOULD YOU PICK

Gotham Knights gives you the option to play through the game as Robin, Batgirl, Red Hood, or Nightwing, but with the freedom of choice comes a difficult question: which character should you pick? Well, the answer to that question is that there is no wrong answer. All four characters are strong in their own specific ways and what differentiates them most is their gameplay style and specialties.

The question of who to pick may seem important during that first decision, but keep in mind that you can switch between characters before going out each night to patrol Gotham's streets. So, instead of giving a straightforward ranking, this guide will give a brief rundown of each character's strengths so that you can choose which skillset sounds best for your preferences.

## Robin

Robin is a stealth specialist who is a perfect choice if you want to take out enemies without ever being seen. He has several skills that help clear rooms as quietly as possible, and they're unique to him. These skills include quiet vantage point takedowns like the ones found in the Arkham games, and the ability to perform silent takedowns on large enemies, among others.

When he does find himself in full-on brawls, he has plenty of ways to create distractions, dip out of combat, and return to the shadows. Robin uses decoys, smoke bombs, and special pellets to avoid direct confrontation.

For the times when he can't get back into stealth, his elemental decoys draw aggro and deal elemental damage at the same time. Robin's ranged attacks also deal elemental damage and can even stick to surfaces to draw attention away from him.

## Batgirl

Batgirl is a technologically savvy hacker who is also surprisingly tanky. She's the right pick if you want a well-rounded fighter who can survive during tough fights, hack cameras for easier stealth, and detonate enemies or technological devices to deal extra damage.

She has skills that give her self-revives, health regeneration, and max HP boosts to keep her in the fight for as long as possible. And when a quieter approach is necessary, Batgirl is able to disable cameras or even become completely invisible to them. That hacking extends to combat to provide AoE damage from tech-related explosions.

If you want to hit harder, her crit chance and crit damage can also be raised, which works well alongside her unique, extended combos that result in plenty of knockdowns.

## Red Hood

Red Hood is more of a ranged fighter who can deal plenty of damage from a distance with the option of grabbing enemies when he gets up close. Choose him if gun combos and strong grabs sound cool to you. At medium or long range, he's able to control crowds while gaining damage multipliers through marksman abilities.

These damage multipliers activate when Red Hood actually aims his guns at enemies instead of performing combos, so you'll want to have a bit of distance between yourself and opponents. But when they do close the gap, his skills allow him to perform grab easily and even turn thrown enemies into AoE bombs, which is where some of that crowd control comes into play.

Red Hood's other passive skills compliment this play style by inducing fear in criminals – a status effect that weakens them through grabs. And if you manage to keep your distance, there are upgrades to shorten his lock-on time allow unlimited gunshots for a limited time.

## Nightwing

Nightwing also excels at crowd control, but from close range. He's the right pick if you want to be agile in combat as you bounce from one enemy to the next while chaining combos. His in-combat mobility is second to none, as he's never in one spot for too long.

Nightwing has skills that allow him to rapidly bounce from one enemy to the next and start evade chains that boost his momentum meter, granting access to special moves more often. These dodges are able to link into AoE attacks that can cause several knockdowns at once, so being surrounded is hardly ever an issue for Nightwing.

Nightwing is also more of a team player than the rest of the group in that he can revive fallen allies in co-op from a distance to make tough team fights more manageable.

As you can see, each of the four main characters has something unique about the way they approach combat. None of them are necessarily outright better than the others, so your choice should be made based on which style of gameplay sounds the most personally enjoyable to you.

You can have a look at each character's skill trees to get a better idea of what they're capable of.

219

# ABOUT THE AUTHOR

When I finding new tricks, tips, and strategies to beat each other, they came up with a brilliant idea. Let's take these hours of gaming expertise, and share these skills with like mind people. At that moment, the Gotham Knights Complete Guide were born. With more exciting gaming books being developed in the Lab as we speak. I am creating a buzz in the gaming guide publishing world, with a ground swell of followers, anxiously awaiting my new releases.